Dragonfly

The Luftwaffe's Experimental Triebflügeljäger Project

David Myhra

Schiffer Military History
Atglen, PA

Book Design by Ian Robertson.

Printed in China.
ISBN: 0-7643-1877-2

We are interested in hearing from authors with book ideas on related topics.

Published by Schiffer Publishing Ltd.
4880 Lower Valley Road
Atglen, PA 19310
Phone: (610) 593-1777
FAX: (610) 593-2002
E-mail: Info@schifferbooks.com.
Visit our web site at: www.schifferbooks.com
Please write for a free catalog.
This book may be purchased from the publisher.
Please include $3.95 postage.
Try your bookstore first.

In Europe, Schiffer books are distributed by:
Bushwood Books
6 Marksbury Avenue
Kew Gardens
Surrey TW9 4JF
England
Phone: 44 (0) 20 8392-8585
FAX: 44 (0) 20 8392-9876
E-mail: Bushwd@aol.com.
Free postage in the UK. Europe: air mail at cost.
Try your bookstore first.

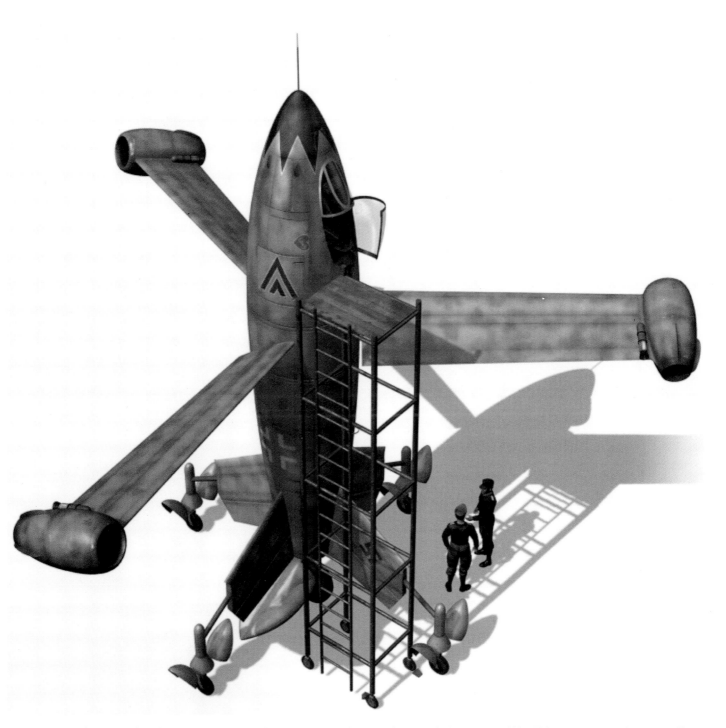

The *Focke-Wulf Fw Triebflügeljäger* project as envisioned by *Dr.-Biology Erich von Holst* in the late 1930s. This proposed vertical take off and landing flying machine is seen from its port side along with a portable metal scaffolding required to allow a pilot to enter and leave its cockpit. Digital image by *Gareth Hector*.

Dragonfly

Introduction

The aeroplane won't amount to a damn until they get a machine that will act like a hummingbird, go straight up, go forward, go backwards, come straight down, and alight like a hummingbird.

- Thomas Alva Edison

This is the story of the world's first attempt at perfecting a true tail-sitting vertical take off and landing (VTOL) interceptor flying machine...*Focke-Wulf Flugzeugbau's* proposed *"Triebflügeljäger"* or thrust wing fighter project of 1944. The *Triebflügeljäger* was not intended to be an air-superiority dog fighter but a bomber killer rising straight up from its hiding place in the forest or urban area to meet and attack Allied bombers head on. With its mission completed it would return to its hiding on the forest floor and wait to arise again to attack another bomber stream. So the *Triebflügeljäger* would have been a point-defense interceptor intended for the *Luftwaffe Home Defense* squadrons and its sole virtue was that it wouldn't need a runway. Highly unusual, then as well as today, the *"Triebflügeljäger"* belongs to a type of aircraft which requires a continuing thrust of air to lift itself straight up and then gradually transitioning to horizontal flight. It does the exact opposite to land vertically. It has been said that more money and effort have gone into the quest for successful VTOL airplanes in the last sixty years and to less effect than can be found in any other branch of aviation research. The main reason for all these design dead ends and failures, it has been stated, is the designer's inability to keep their VTOL flying machine from becoming too complicated. It would have been equally difficult if not more so for the *"Triebflügeljäger's"* planners and designers. Some of the best aeronautical minds of their generation turned their attention to this project. These individuals included zoologist *Erich von Holst*, aerodynamicist *Hans Multhopp*, *Flugbaumeister Heinz von Halem*, gas dynamitist *Otto Pabst*, engineer *Hermann Langner*, and others struggled with the challenges a true VTOL flying machine presented. Aside from the pilot's severe challenge to vertically a set a VTOL machine back on the ground without visually seeing it, *Otto Muck* of *Junkers Flugzeugbau* suggested in 1938, a VTOL with a large diameter double propeller and a telescoping aft fuselage to allow the pilot a better view of the ground because the tail assembly moved closer to the wing's trailing edge. He obtained a patent for such a machine on September 10th 1938. Lifting the estimated 11,000 pound ma-

chine would have required enormous quantities of horsepower. It was *Hans Multhopp* who suggested that the necessary horsepower come from small diameter ramjet engines, one each at the three rotor tips based on the promising research by rocketeer *Eugen Sänger* at *DFS*-Ainring. Looking back to those days from the advantage of the present, it is doubtful that a ramjet power unit would have been practical. In essence, the ramjet is nothing more than an open cylinder with an internal fuel spray. The device requires a minimum air flow velocity of about 300 miles/hour to become effective and this is usually obtained by means of a solid or liquid propellant booster. At about 300 miles/hour the air-fuel mixture is

A ground level view of the *Focke-Wulf Fw Triebflügeljäger* as seen from its starboard side. The pilot is seated within the vertical take off flying machine. The pilot's entry and egress scaffolding has been moved away as the pilot makes preparations to begin the rotation of its three rotors. Digital image by *Gareth Hector*.

ignited and burns continuously as long as the air stream is passing through the duct at that minimal rate of flow, the gases emerging from the exit with a greater energy than that of the air-stream entering the intake. This differential of exit and entry energy, resulting in a higher exit velocity, imparts a very strong thrust to the duct by the principle of reaction.

The well known problem of maintaining ramjet efficiency at all altitudes of operation would have plagued *Otto Pabst*, *Focke-Wulf Flugzeugbau's* expect on the dynamics of gases. He was taking *Eugen Sänger's* basic ramjet research and studying ways to make effective small diameter short length ramjet engines. In short, a compact ramjet engine. The ramjet engine he was seeking to perfect would have an operational ceiling of about 60,000 feet. Furthermore, the ramjet, which really only becomes practical and economical at high sonic velocities of around Mach 2.0 (1,483 miles/hour), is a known fuel guzzler below Mach 1.0 (741 miles/hour) consuming fuel at an astonishing rate second only to bi-fuel liquid rocket engine such as *Wernher von Braun's A-4 (V2)* rocket missile of that era. Would *Otto Pabst's* research made it more thrifty? Did our *Triebflügeljäger* designers and engineers realize these challenges? Did *Hans Multhopp* when he decided that the basic ramjet engine research of *Eugen Sänger* might make for a practical engine for a VTOL? Or were they all caught up in the ramjet's untested ease of installation, serviceability, reliability, low cost, economy of manufacture, and promise of high horsepower? It is unclear. Despite the ramjet's apparent simplicity, it presented itself with some formidable design problems as rocketeer *Eugen Sänger* was experiencing in his ramjet research and experiments at the *Deutsche Forschungsanstalt für Segelflug*-Ainring (*DFS*) or the German

Research Institute for Sailplane Research at Ainring. In that time and place the ramjet and its burners were no more simpler in essence than the struggles faced by rocketeer *Walter Thiel* of Peenemünde who at the time was designing the *A-4's (V2)* pure rocket combustion chamber with its 18 injectors. Both had expulsion systems which relied on the complexity of producing pressure. *Triebflügeljäger's* aircraft fuselage and ramjet engine designers may have not yet realized that the ramjet engine would have been a poor choice just as it would be today for a VTOL. With all this said, lets continue on with the amazing story of men who wanted a flying machine capable of take off, hovering, forward flight, and landing as does the common dragonfly.

The concept for a VTOL did not originate within *Nazi* Germany's conventional aircraft design community. The idea came instead from a professor of zoology at the Göttingen Technical University by the name of *Erich von Holst* (1908-1962). In fact it was *Professor von Holst* who coined the word "*Triebflügel*" based on his extensive research into bird and insect flight. He was especially inspired by the strongly flying common dragonfly (genus Cordulegaste with between 400 and 500 species in the United States alone) with its ability to fly vertically, horizontally, hover, backward , and sideward. Its maximum speed is between 25-30 miles/hour and its cruising speed is about 10 miles/hour. It was *von Holst's* belief that the principles by which dragonflies took flight could be applied to a man made flying machine. This, then, is the story of the dragonfly which inspired a VTOL. But first, a word about *Professor Erich von Holst*. *R.D. Martin*, translator of his seminal research work *Zur Verhaltens-physiologie bei Tieren und Menschen* or *The Behavioural Physiology of Animals and Man*, Volume #1, said this about *Erich von Holst*:

Focke-Wolf Fw Triebflügeljäger's **rotor blades are in motion thanks to a** *HWK* **bi-liquid rocket engine mounted at each wing tip inboard the** *Pabst* **ramjet engine. The rotor blades will have to rotate at approximately 250 to 300 miles/hour in order for the ramjet engines to become self sustaining. Digital image by** *Gareth Hector*.

Its vertical lift off complete, a *Focke-Wulf Fw Triebflügeljäger* joined up with a *Messerschmitt Me 262 A-1a* turbojet powered fighter in the hunt for American high flying *Boeing B-17* bombers. Digital image by *Gareth Hector*.

First, *Erich von Holst* is considered one of those rare individuals whose creative activities merit the term "genius" without a shadow of doubt. One is struck by two major features of his work which unerringly evoke admiration he wrote. In the first place, *von Holst* had a unique aptitude for remorseless adherence to scientific method in pioneering studies, and yet managed to combine this with a written style of amazing simplicity and clarity. The second remarkable feature of *von Holst's* investigations is his breath of coverage. It is, at first sight, almost inconceivable that a single man could master so many different areas of investigation in one lifetime, particularly since each study was marked with a special stamp of originality which unfailingly threw fresh light on the field involved.

Dr. Erich von Holst's "*Triebflügeljäger*," as it came to be seen on design drawings, received impute from gifted aerodynamicist *Hans Multhopp*. Design of the *Triebflügeljäger's* air frame and component placement was turned over to *Flugbaumeister Heinz von Halem* and he was supported by *Ludwig Mittelhüber* and *Dipl.-Ing.*

***Focke-Wulf Fw Triebflügeljäger* is seen with two other companions streaking up to confront American *Boeing B-17* bombers. Digital image by *Gareth Hector*.**

Dragonfly: the Luftwaffe's Experimental Triebflügeljäger Project

Quenzer. Ludwig Mittelhüber and *Kurt Tank* had both joined *Focke-Wulf* together both coming from *Messerschmitt AG* in 1930. *von Halem* had joined the *Focke- Wulf Flugzeugbau* about mid 1943. Although *von Halem's* design work was finished in September 1944, the *Otto Pabst* ramjet engines which had been recommended by *Hans Multhopp* and which had been under development at *Focke-Wulf Flugzeugbau* since 1942, never did become operational due to a lack of time and resources prior to war's end. The *Triebflügeljäger* would have had an overall height of 30 feet [9.15 meters] as it stood upright on its tail, its fuselage would be surrounded about its mid-section by three 30.2 foot [9.2 meter] long rotors, which at each tip was a 2 foot 3 inch [68.58 centimeter] diameter *Otto Pabst* ramjet engine. Once these ramjets were ignited, *Triebflügeljäger's* three rotating arms...performing like blades of a helicopter...would lift the flying machine off the ground and take it quickly aloft in hot pursuit of high-flying American *Boeing B-17* four engine bombers. Although the *Triebflügeljäger* was not given a designation number by the *Reichs Luftfahrt Ministerim* (*RLM*) or German Air Ministry, without a doubt this design by *Focke- Wulf Flugzeugbau* has to take top honors for being one of

A rear starboard side view of *Professor Erich von Holst's* "*Schwingenflugmodell I*" with its bird-like horizontal stabilizer.

the most unusual fighter interceptors under under consideration during WWII. In addition, *Erich von Holst's Triebflügeljäger* can be considered as the forerunner of all modern vertical take-off flying machines.

The notion of a VTOL was the subject of a patent (patent number 728 256) granted on September 10th 1938 to the German engineer from *Junkers Flugzeugbau Dipl.-Ing. Otto Muck*. However, nothing ever came of it. Seventeen years prior, about 1921, rocketeer *Dipl.-Ing. Hermann Langner* wrote and illustrated a two part article which appeared the German magazine *Der Luftweg*. The title of *Langner's* article was *"Die geschichte einer Mars-expedition"* or The Story of a Mars Expedition and it describes a future trip to Mars in a giant spaceship, the *Meteor VI*. The spaceship is shaped like a spindle with a very long tail, tapering to a needle-like point, broken only by the exhaust of one of its three rocket engines. However, about one third the way down from the *Meteor VI's* nose is what appears to be a very large two-bladed propeller attached to a collar similar to how the three rotor blades were to have been attached on the *Triebflügeljäger*. About amidships on *Langner's Meteor VI* are a pair of short, broad wings. Additional rocket engine's thrust project at an angle just behind the rotating propeller. The very larger rotating propeller coupled with rocket engines, *Langner* believed, would allow his spaceship to take off vertically from Earth without the aid of any launch tower. The rotating propeller would allow the *Meteor VI* to vertically land on Mars, too. It appears that *Hermann Langner* was a consultant to the *Triebflügeljäger* team. It would be *Focke-Wulf Flugzeugbau's* director *Kurt Tank* who would have his aeroengineers, especially *Hans Multhopp* take *Professor von Holst's* idea and pursue it further as we see in their *Triebflügeljäger* proposal. It appears that *Kurt Tank* became interested in von *Holst's Triebflügel* after he and *Dr.-Ing. Eugen Sänger* from (*DFS*) met in early 1943 at *Focke-Wulf's* offices at Bad Eilson. The relationship between the rocketeer *Sänger*, the zoologist *Erich von Holst*, and his *Triebflügel* is not clear although it may have

Professor Dr.-Zoology Erich von Holst (**1908-1962**) **seen at this desk at the Göttingen Technical University about 1938. Above the thirty year old Professor and to the right can be faintly seen a paper and wooden stick model of his "*Schwingenflugmodell I*" hanging from the ceiling.**

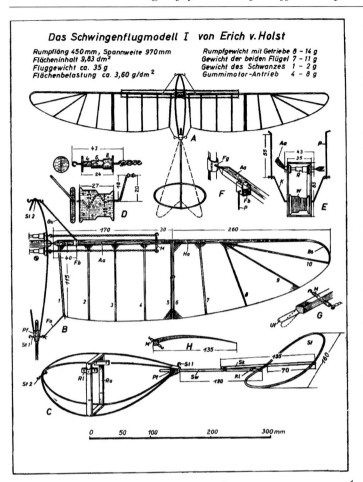

Das Schwingenflugmodell I von Erich v. Holst

Rumpflöng 450 mm, Spannweite 970 mm
Flächeninhalt 9,83 dm²
Fluggewicht ca. 35 g
Flächenbelastung ca. 3,60 g/dm²

Rumpfgewicht mit Getriebe 8 – 14 g
Gewicht der beiden Flügel 7 – 11 g
Gewicht des Schwanzes 1 – 2 g
Gummimotor-Antrieb 4 – 8 g

come about due to *von Holst's* article published in the 1942 Aeronautics Research Annual. Later on during the war about mid 1944 to early 1945, *Eugen Sänger* was involved in *Ernst Heinkel's* proposed VTOL *Wespe* or wasp project. He is shown in a photograph from that time along with *Professors Ernst Heinkel, George Madelung, Heinrich Focke, Dipl.-Ing. Siegfried Günter* and *Dr.-Ing. Arthur Weiss*. *Hans Multhopp*, the brilliant young aerodynamicist, had come to *Focke-Wulf Flugzeugbau* in 1938 from *Aerodynamische Versuchsanstalt (AVA)*, appeared to fully believe the concept was feasible. Since a VTOL can lift its own weight, it could be built in any shape, *Hans Multhopp* believed. From mid 1943 and on, it appears that *Focke-Wulf Flugzeugbau* were seriously studying the project's feasibility and making aerodynamic tests with the assistance of *Professor Adolf Busemann*, head of (*LFA*) Braunschweig- Volkenrode. *Professor Walter Georgii, Eugen Sänger's* boss at *DFS*, was also chairman of the all powerful four man *Reich* Research Council (others included *Professor Ludwig Prandlt* of Göttingen Technical University, *F. Seewald* of Aachen, and *Adolf Beaumker*) at this time and any research carried out at the several aviation research institutes in *Nazi* Germany such *AVA* and *LFA* and others, had to have prior authorization of the *Reich* Research Council which *Georgii* chaired. Due to *Georgii's* blessing, the *Triebflügeljäger* project appears to have had full cooperation among the aviation research institutes. As a result, the *Triebflügeljäger* project could not have been merely some whimsical idea but a very serious academic endeavor which was being

Above: Pen and ink three-view drawings of *Professor Erich von Holst's* "Schwingenflugmodell I" featuring the internal structure and shown in millimeters.

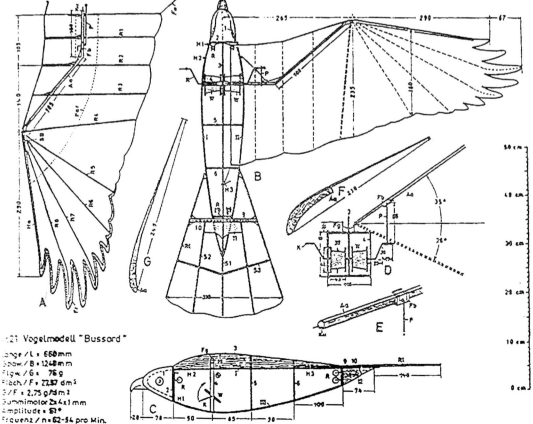

Right: Another bird-like model by *Erich von Holst* which he called the "*Bussard*" or buzzard. It had a wing span of 1240 millimeters [4 feet]. Notice the bird feather-type wing tips.

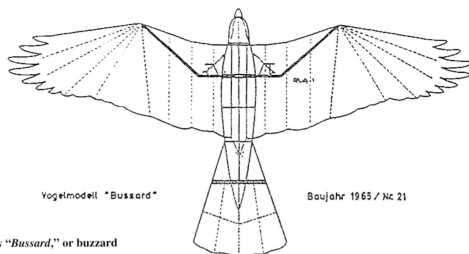

Vogelmodell "Bussard"   Baujahr 1965 / Nr. 21

A pen and ink illustration of *Erich von Holst's* "*Bussard*," or buzzard bird, as seen from above.

supported by highest ranks including *Oberst Siegfried Knemeyer* of the *RLM* and *Professor Walter Georgii* of the *Reich* Research Council. It appears, too, that *Eugen Sänger*, rocketeer, fully cooperated with *Focke-Wulf Flugzeugbau*, especially *Dr.-Ing. Otto Pabst*, their expert on gas dynamics, by providing his substantial knowledge of *Lorin* duct engine. To help *Heinz von Holst's* team move along a little faster, a new engineer came to *Focke-Wulf Flugzeugbau* in mid 1944 by the name of *Major Schroedter*. Cooperation between *Professor Busemann* and *Dr.-Ing. Theodor Zobel* of *LFA*-Volkenrode by Braunschweig, *Professors E. Wolfhand Schmidt* and *Albert Betz* of *AVA* -Göttingen, and *Focke-Wulf Flugzeugbau* continued to flourish. *Busemann* preferred to research the basics of gas dynamics while *Schmidt* treated mainly physical and chemical questions in engine combustion. Throughout this intense period of activity, zoologists *Dr. Erich von Holst* and *Dr. D. Küchemann* continued to be involved in the *Triebflügeljäger* project and it appears that both naturalists visited frequently the Bad Eilson offices of *Focke-Wulf Flugzeugbau*. Although this illustrated history features the *Focke-Wulf Triebflügeljäger*, German aircraft designers proposed several other competing vertical take off and landing aircraft: These flying machines' different shapes and different points of view included:

• *Bachem Ba 349* "Natter;"
• *Fieseler Fi 166 Projekt*;
• *Focke-Angelis Fa 269*;
• *Focke-Wulf* VTOL *Projekt*;
• *Heinkel He Projekt 1077* "Julia;"
• *Heinkel He Projekt* "Lerche II;"
• *Heinkel He Projekt* "Wespe;"
• *Junkers Ju EF-009 Projekt*;
• *Von Braun* VTOL *Projekt*;
• *Von Braun* piloted *A-9* "Manned Reconnaissance rocket missile" *Projekt*;
• *Weserflug Projekt 1003*;

Erich von Holst Zur Verhaltensphysiologie bei Tieren und Menschen

Gesammelte Abhandlungen Band II

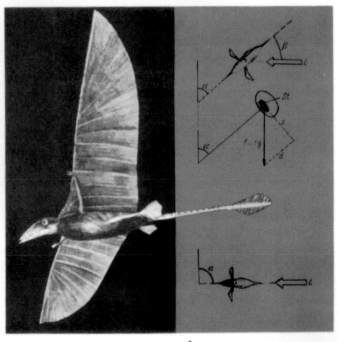

piper

A photo copy of the dust cover jacket from *Professor Erich von Holst's* book "*Zur Verhaltens-Physioilogie bei Tieren und Menschen*" or "*The Behavioural Physiology of Animals and Man*." The dust jacket features one of *von Holst's* paper and wood stick birds known as the "*Rhamphorhynodus*."

A nose starboard side view of *von Holst's "Schwingenflugmodell II"* in flight in the late 1930s.

***Erich von Holst's "Schwingenflugmodell II"* in flight and appearing bird-like in its form and shape.**

The conceptual idea of a VTOL aircraft in Germany at least, began with the studies in the late 1930s by *Professor Dr. Erich von Holst* (1908-1962), a zoologist/naturalist from the Göttingen Technical University. He viewed that the flight of birds and insects could be looked upon as a possible source of inspiration from which hu-

man engineering could profitably learn and copy. *Von Holst* was aided by two other colleagues: *Professor Dr. Dietrich Küchemann* and *Professor Dr. K. Solf* (members of the *Aerodynamic Versuchsanstalt zu* Göttingen (*AVA*) or the Aerodynamic Research Establishment at Göttingen). It appears that these three academics

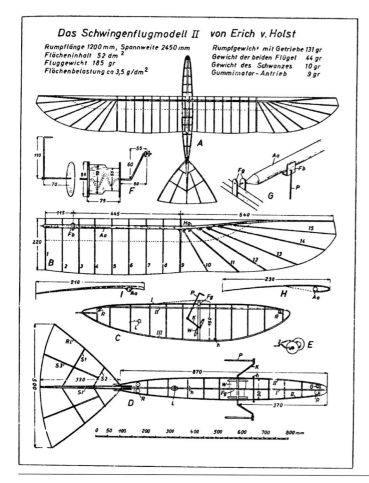

Another view of *Erich von Holst's "Schwan"* as seen from its nose port side. *Professor von Holst* appears in the lower right hand corner of the photograph.

A pen and ink three view drawing of *von Holst's "Schwingenflugmodell II."* It was also known as the *"Schwan,"* or swan.

had for a long time, been curious about the mechanics of bird and insect flight, and *von Holst* hoped that he might be able to apply some of his research findings to future German aircraft design. Of particular interest to these men was a bird's ability to create enough lift for flight regardless of how fast it was going to fly once it got airborne. Birds, *von Holst* observed, achieved lift through the flapping motion of their wings. This was not a technically desirable solution in aircraft because of the very high force developed by such slow up-and-down motion, but there was one flying creature that achieved lift without flapping its wings...the well known dragonfly. This large insect, he noted, used two sets of wings, one behind the other, with the sets swinging 180 degrees out of phase, that is, one set going up while the other was going down.

Erich von Holst wrote that there were methods and devices employed by the flying animal which could be usefully applied in aeronautical engineering, it must not be overlooked that in many cases, their utilization could not be a matter of mere copying but

This is the common "dragonfly" with its two pair of wings. It was the dragonfly's natural winged flight which *Professor Erich von Holst* was attempting to mechanically duplicate with his *triebflügel.*

rather of suitable adaptation to engineering needs. He believed, for instance, that it was clear that no striking aerodynamic benefit could be derived from the flapping motion of birds. So any engineering application therefore, would have to be transformed into a rotation about an axis. In this light *von Holst* viewed various forms of cyclogiro rotating-wing systems such as the cycloidal *Voith-Schneider* propeller (1933) with its controllable blade angles which had already been applied to ship propulsion. To *von Holst*, even the propeller was the successful development of the functions of the outer part of a bird's wing. He found that there was one general principle which was almost universally employed by the flying animals: lift and thrust were generated simultaneously by the same organ; and there was no question of the addition of a propulsion unit to the lifting surface. *von Holst* believed that it was possible that this principle might also be applied in the future course of aeronautical engineering. One, way, for example, would be the complete integration of turbojet engines into the wing. Or, alternatively, at the other extreme, another way would be to dispose of the wings altogether and leave only the propeller, which would produce the necessary lift by operating with a small inclination to the direction of flight. *Professor von Holst* came to this conclusion with the help of the brilliant aerodynamicist *Hans Multhopp.* For example, beginning with the dragonfly, the first step would be to make the wings rigid and to add a tail plane, replacing the long body, which is otherwise needed to counteract the large alternations in pitching moment caused by the periodic changes of the lift forces on the two wings. The next step indicated would automatically eliminate the pitching moment variations, but only the final transition to counter rotating wing (or propellers) is of real engineering significance.

In 1940 *Erich von Holst* participated at the *Saalflugwettbewerb* or Indoor Flying Scale Model Competitions at Breslau. During the competitions he demonstrated a flying model of his

Professor Konrad Lorenz (left) and *Professor Erich von Holst* (right). These two highly respected men from Göttingen University were photographed together in the mid-1950s.

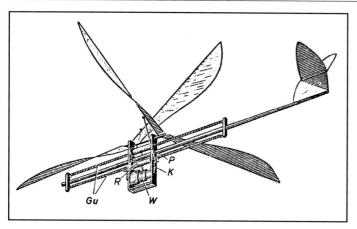

This is a pen and ink drawing by *Professor von Holst* in his attempt to duplicate the dragonfly and its twin pair of flapping wings.

Gu - **rubber band motor - two**	R - **thread roll**
K - **crank**	W - **stepped roll plate**
P - **connecting rod**	

The first example of *Professor von Holst's* wood stick and paper dragonfly flying model with its twin pair of flapping wings.

Schwingenflugzeug or ornithopter, flapping wing design, which he called his *"Libelle"* or Dragonfly. He also experimented with several different versions and eventually, he and two of his fellow professors published their findings in the prestigious *Jahrbuch der Luftfahrtforschung* 1942 or the Aeronautical Research Annual of 1942. The title of the report was simply *"Die Triebflügel."* The authors put forth the hypothesis that a flying device could be constructed which would allow vertical lift-off transitioning to horizontal forward flight, and then a vertical controlled descent/landing. *von Holst's* flying models were driven by small rubber-band motors of his own design, and the ratio of the model's weight to the horsepower of the motor was usually very high (2 pounds per horse-

power). The motor's power might seem low compared with what an aircraft requires, but if the laws of similarity are considered it would be seen that this was not so but in fact a very powerful engine. In the colorful language of modern aeronautics, *Focke-Wulf's projekt "Triebflügeljäger"* would be called a vertical take-off and landing (VTOL) aircraft. The three authors believed that since it could takeoff vertically takeoff machine would be superior to any of the *Luftwaffe's* fighter/interceptors which required long runways for takeoff and landing which at this time of the war were being heavily bombed by Allied bombers. Not so with the *Triebflügeljäger* which could takeoff and land on any flat surface...among the trees, back streets, farm yards, city parks, and so on.

Professor von Holst's **second version of his mechanical vertical take off dragonfly, or *Triebflügel*, meaning driving wing. This version by *von Holst* no longer has the flapping motion of the common dragonfly; instead *von Holst* has changed the wings to appear more like a three-bladed propeller and counter rotating. It is seen in flight. In addition, *von Holst* moved the rotors to the tip of the wood stick fuselage.**

Dragonfly: the Luftwaffe's Experimental Triebflügeljäger Project

The idea for a VTOL flying machine was becoming increasing popular due to *Nazi* Germany's growing inability to launch conventional fighters such as the *Messerschmitt Me 262 A-1a* was that when landing they required cover/protection of *Focke-Wulf Fw 190* fighters to keep the valuable jet fighters from being shot out of the sky. The new high speed turbojet powered fighter required a long slow landing approach. This was being taken advantage of by free-roaming long-range *North American P-51* fighters which often shot *Messerschmitt Me 262s* down as they started the landing approach. Thus, the *Triebflügeljager projekt*, as well as *Erich Bachem's Ba 349 "Natter,"* a vertical take off rocket interceptor, were ideas necessitated by dire circumstances of the enemy (America) gaining air superiority over the German homeland. VTOL projects were being strongly pushed by the *RLM's* head of Technical Development *Oberst Siegfried Knemeyer*. He, too, believed as did *Hans Multhopp*, that since a VTOL can lift its own weight, it could be built in any shape.

Professor von Holst went on to build a scale model of his *Libelle* he found that he could imitate a dragonfly's flight characteristics. But he also found the two complete sets of movable wings a bit too complex, and he began to simply his artificial dragonfly. First, he changed to two rigid wings, one behind the other, which moved about hinges parallel to the body axis, again 180 degrees out of phase. His next step in simplifying the dragonfly's flight mechanics was to replace the two wings with the technically much simpler

complete rotation about the same fuselage axis. After some experimentation with the size and location of tail surfaces, *von Holst* came up with a model flapping wing airplane design that flew any direction in space - vertical or horizontal - or at any odd flight path angle, with a contra-rotating propellers acting as propeller and lifting device.

Erich von Holst discovered another general principle from his study of birds and that was that the flight of animals was always to have the plane of the transverse oscillation nearly normal to the direction of motion. This meant that the axis of the thrust wing or *Triebflügeljäger*, as the final stage may be called, should likewise be nearly in the direction of flight. The flying machine would take off with its axis vertical, resembling a helicopter; but unlike the helicopter, it would not keep its axis vertical but turn over gradually, until in high-speed flight it would be almost horizontal. However, it is not clearly understood if the *Triebflügeljäger*, after it turned horizontal, would have continued on with its rotors spinning as they had during take off. We have to assume that the *Triebflügeljäger* would have taken off like a helicopter and cruised like a conventional winged aircraft. Furthermore, it is not clear how the *Triebflügeljäger* would have tackled the mechanics of switching from rotor to wing-borne flight if in fact this was how it would have done. In the crucial moments of transition when the rotor is slowing down on its way to becoming a wing the *Triebflügeljäger* could have lost lift and potentially fallen out of the sky.

A photo of *Professor Erich von Holst's* wood stick and paper dragonfly, or *Triebflügel*, with its counter rotating propeller-like blades and shown in vertical flight.

A poor quality photo of *Professor Erich von Holst's Triebflügel*, a wood stick and paper version of the common dragonfly as seen in vertical flight shortly after lift off.

For example, once the *Hawker-Siddeley Aviation's Harrier* "Jump Jet" was off the ground vertically, wrote *Bruce Myles* in his book *Jump Jet: The Revolutionary V/STOL Fighter*, and the pilot controlling its flight path away from the ground, he or she then started to rotate the turbojet engine's nozzles aft. There were limits to how quickly this could be done. If the pilot was sitting at 50 feet and was foolish enough to slam the nozzles rearwards, which could be done in about one third of a second, then the laws of *Sir Isaac Newton* would dominate. The airplane would drop like a stone. Instead the pilot rotates the nozzles initially around ten to twenty degrees from the vertical.

Furthermore, the *Triebflügeljäger* would have needed a variable pitch rotor, a dangerous thing and keeping it from self-destructing is a real challenge. On the other hand, the rotor on a traditional helicopter spins at a constant high speed to minimize vibration...about 450 to 500 revolutions per minute. When a conventional helicopter is moving slowly, its rotor is turning much faster than necessary wasting a great deal of power. So would the *Triebflügeljäger* had a variable speed rotor? It is unclear. The reason a variable speed rotor has rarely been attempted is that at intermediate speeds the flying machine can vibrate so intensely that it self destructs.

Above: A poor quality photo of the wood stick and paper *Triebflügel* seen in horizontal flight. It has a "T" tail favored by *Focke-Wulf's* brilliant young aerodynamicist *Hans Multhopp*. Right: *Professor von Holst* experimented with different forms and positions of mechanical wings for his Dragonfly. This pen and ink illustration features the evolutionary steps *von Holst* went through, beginning with the common dragonfly to his final design with the contra rotating rotors on the final model.

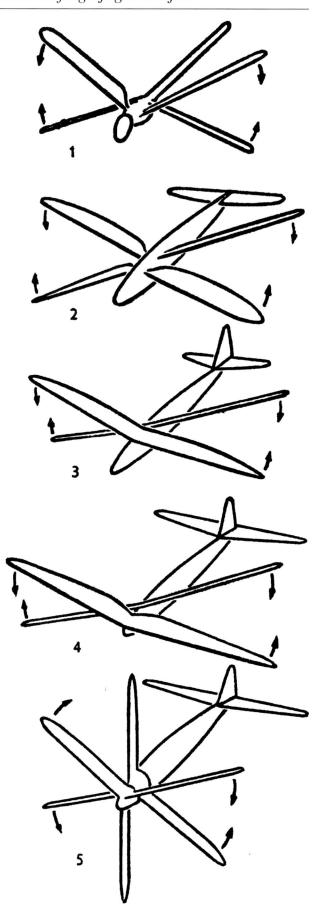

Digital artist *Gareth Hector* has recreated the evolutionary steps taken by *Professor Erich von Holst* in his quest to achieve the same capabilities as a common dragonfly. Featured in this digital image is the common dragonfly, the starting off point for *Professor von Holst* to perfect, mechanically, the ability of the dragonfly to lift off vertically, fly horizontally, and land vertically.

In *Gareth Hector's* third image, *von Holst*, in his model, has moved the contra flapping wings to the top (dorsal) position on the fuselage. In addition, *von Holst* had added a vertical stabilizer (rudder) to the tail assembly.

We do not know how *Erich von Holst, Hans Multhopp, Heinz von Halem,* and *Ludwig Mittelhüber* would have had the *Triebflügeljäger* make the transition to horizontal flight from vertical, that is, overcoming the flight-switching problem, if in fact, the machine was designed to do that. We do not know for sure. However, to do so the *Triebflügeljäger* would might have required some type of lifting surface other than its rotors to tide it over the transition. It might have required a forewing or front wing (canard) in addition to its horizontal stabilizer at its tail something big enough to lift the entire weight of the *Triebflügeljäger,* perhaps 11,000 pounds including armament, during the transition between vertical take off and airplane modes. This would have taken the lift loads off its three rotors (each rotor lifting approximately 3,700 pounds in addition to substantial dynamic loads as well), making it easier

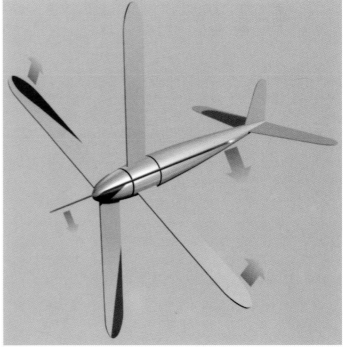

Digital artist *Gareth Hector's* second image featuring *Professsor von Holst's* evolutionary transformation of the common dragonfly into the *Triebflügel. Von Holst* has substituted the long cylindrical body of the dragonfly for a fuselage. Although he still retains the contra flapping wings of the dragonfly, *von Holst* has added a horizontal stabilizer to the tail of the fuselage.

In *Gareth Hector's* final digital image featuring the evolutionary *Triebflügel, von Holst* has moved the wings to the nose of the fuselage. Instead of flapping as does the dragonfly, the two pair of wings now contra rotate on the fuselage.

to slow down and stop all the while the flying machine has sufficient air speed to keep the *Otto Pabst* small diameter short length ramjet engines providing sufficient thrust for propelling the *Trieflügeljäger* forward. The well-known speed limitation of the helicopter would thereby be overcome by *Professor von Holst's* thrust wing.

All rotor systems are subject to asymmetry of lift in forward flight. It is not clear how *von Holst* was going to handle this situation. For example, the *Triebflügeljäger* while hovering its lift on the three rotors would be equal across the entire rotor disk. As the *Triebflügeljäger* gained air speed, the advancing rotor would have developed greater lift because of the increased airspeed and the retreating rotor would have produced less lift. This asymmetry might have caused the *Triebflügeljäger* to roll. For example, if its rotor speed equaled 249 miles/hour [400 kilometer/hour] and it was moving forward at 62 miles/hour [100 kilometer/hour] then the advancing rotor would have a speed equal to 311 miles/hour [500 kilometers/hour] but the retreating rotor would have a moving speed of only 186 miles/hour [300 kilometers/hour]. This dissymmetry would have had to been compensated for in some way.

Professor von Holst's scale model of his ornithopter or flapping wing was tested at the *AVA- Göttingen* which was the premier center for basic aerodynamic research in Germany. It was here that

Ludwig Prandtl, Albert Betz, Dieter Küchemann, K. Solf, and others were able to make those important contributions to aerodynamic theory which won for Göttingen its international reputation. So, with *Erich von Holst's* ornithopter, it was found that it did not compare unfavorably with orthodox airplanes of the same era. However, significant technical problems remained, particularly the mechanics of the large contra-rotating propeller and location on the fuselage, a gearbox, and the power plant with a huge horsepower output. *AVA*-Göttingen engineers estimated that such an aircraft like the proposed *Focke-Wulf Fw Triebflügeljäger* weighing in at about 10-ton would require 7,500 horsepower, which at that time in the early 1940s, was beyond the state-of-the-art in Germany. Thus a third step - finding a suitable power plant - was absolutely necessary. The solution came from a different direction. *Focke-Wulf Flugzeugbau* had recently employed the aerodynamicist *Hans Multhopp.* He left *Professor Prandtl* and his Ph.D. studies at the Technical University of Göttingen and joined noted aircraft designer *Kurt Tank. Hans Multhopp* never did finish his Ph.D program. It was *Multhopp* who, working with *von Holst* in 1944, prepared a detailed design which suggested the use of *Otto Pabst* ramjet engines to drive each blade individually and therefore make it possible to dispose of the second system of blades without introducing a reaction moment about the fuselage. *Multhopp* found that the ramjet engine, which was under development by *Otto Pabst* of *Focke-Wulf Flugzeugbau,* would be a particularly suitable application of the ramjet, since its velocity can be kept constant, and it can always work at the design point with the optimum efficiency. In addition, the revolutions per minute (rpm) of the blades would thus be highest at take off and landing, and lowest at top speed. *Multhopp* suggested that the main flight control of *von Holst's Triebflügeljäger* would come about by changing the pitch of the blades. The result, said *Multhopp,* would be an aircraft which combined the high top speeds attainable by conventional modern aircraft of the day with the maneuverability, the ability to climb in any direction, and the

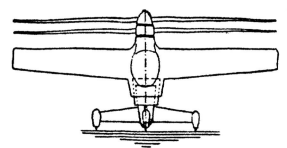

b29 01 Pat. 728 256 v. 10. 9. 38, veröff. 24. 11. 42. Dipl.-Ing. Otto Muck, Uffing. *Flugzeug mit auch senkrechtes Starten. und Landen ermöglichendem Schraubenvortrieb.* Patentansprüche

1. Flugzeug mit auch senkrechtes Starten und Landen ermöglichendem Schraubenvortrieb und einer am Rumpfeck vorgesehenen Standvorrichtung, dadurch gekennzeichnet, daß der Rumpf teleskopartig

in sich verschiebbar gemacht und mit an sich bekannten Öl-, Luft oder Federdämpfungen zur Aufnahme des Landungsstoßes versehen ist.

2. Flugzeug nach Anspruch 1 mit Schwanzleitwerk, u. durch gekennzeichnet, daß das Steuergestänge für das Leitwerk aus zwei verdrehbaren, in sich gemein im mit dem Rumpfe teleskopartig verschiebbaren gegebenenfalls zur Abfederung des Rumpfes m t heranziehbaren Rohren erfolgt.

In the late 1930s individuals in Germany other than *Professor von Holst* were considering ways to lift an air frame vertically off the ground. The only practical method in the late 1930s was the use of ultra large diameter propellers. This pen and ink illustration from the 1930s shows the evolutionary thinking of applying a propeller for a vertical take off flying machine and the final elimination of the wing. Initially aero engineers began their thinking with a conventional sized propeller with power supplied by a conventional piston motor in the nose and wing span. A conventional propeller would not have been adequate. Aero engineers then thought of increasing the diameter, increasing the lifting area of the single propeller, which would have required an enormous increase in horsepower. Seeking more lifting area caused aero engineers to experiment with double or dual propellers rotating counterclockwise to each other. Notice, too, the decreasing span of the wing. A dual propeller arrangement would have taken up almost all the space in the nose of the vertical take off flying machine. As the evolutionary arrangement for a dual propeller powered vertical take off flying machine continued, designers located the propellers aft of the nose; the propellers, now having the diameter of the aircraft's former wing span, thus entirely eliminated the wing. The pilot's position, it is assumed, would have been aft of the dual contra rotating propellers about midship.

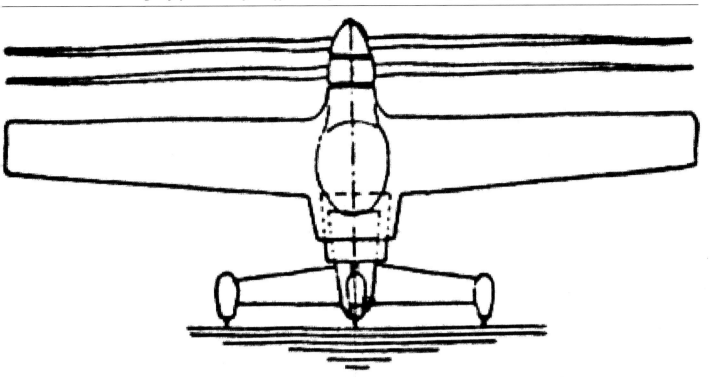

One aero engineer in Germany in the late 1930s applied for and received a patent for a vertical take off flying machine featuring dual contra rotating propellers. *Dipl.-Ing. Otto Muck* proposed a vertical take off flying machine with both dual propellers and wings. Featured is a pen and ink drawing accompanying the patent taken out by *Dipl.-Ing. Otto Muck* in September 10th, 1938. *Muck* wrote that his vertical take off flying machine was to have been powered by a "screw propeller drive system." The text of the patent says that this double screw (propeller) drive system enables this proposed flying machine to start and land vertically. The tail assembly, according to the patent papers says that this flying machine's tail assembly serves also as a support stand. Finally, the patent paper says that the fuselage of this vertical take off and landing machine telescopes in and out, allowing easy transport.

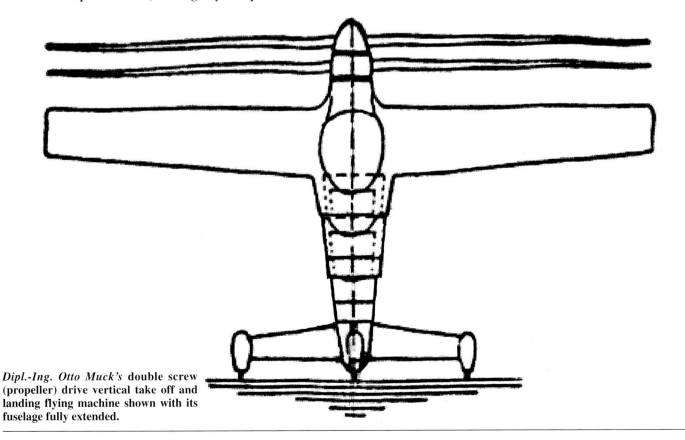

Dipl.-Ing. Otto Muck's **double screw (propeller) drive vertical take off and landing flying machine shown with its fuselage fully extended.**

Though *Professor von Holst's* experiments with a wingless vertical take off flying machine powered by double contra rotating propellers came to be the accepted arrangement, *Nazi* Germany had no aero engine with the required horsepower transfer gear box to power the proposed double contra rotating propellers. It was the *Focke-Wulf* aerodynamicist *Hans Multhopp* who suggested that *von Holst* abandon the double contra rotating propellers. Instead, *Multhopp* suggested three rotors powered by miniature turbojet engines, or ramjet engines. These engines would be mounted on the rotor tips, thus doing away with a huge piston aero engine located in the fuselage, along with its complex power transfer mechanism.

ability to rise into the air without a take off run, which so far, said *von Holst*, only the birds had mastered. Normally a helicopter rotor in hover is a propeller at zero advance; a rotor in ascent/descent has a positive or negative advance rotor; a rotor in forward flight is intrinsically unbalanced due to a rolling moment created by the asymmetry of the loading. It appears that the three rotors of *von Holst's Triebflügeljäger* would not have operated in the several different states and speeds of a typical helicopter. But it might have been a more complicated airscrew or propeller of all.

Focke-Wulf Flugzeugbau had been interested in developing turbojet engines for its own use. However, the *RLM* , under pressure from engine makers such as *Junkers-Jumo, Heinkel AG,* and *BMW*, had forbidden *Focke-Wulf Flugzeugbau* from manufacturing their own turbojets, with the exception of ramjet engines.

When *Focke-Wulf's Dr.-Ing. Otto Pabst* was working on ramjet engines in the early 1940s along with wind tunnel expert *Dr.-Ing. Theodor Zobel,* from *LFA* at Braunschweig, practical application for their use were not readily apparent.

Ramjets are by far, the simplest air breathing aero engines available. Whereas piston engines and turbojet engines use complex mechanical components to compress the air, add fuel, and then extract the energy for useful work, the ramjet with no moving parts whatsoever, compresses the air, adds fuel, and expands the exhaust simply by means of it's shape and its continuous movement through the air. Simple ramjets by *(Roy) Marquardt Aircraft* have been operated at speeds as low as 250 miles/hour when tested on a *Bell Aircraft P.83* in March 1947 and as fast as 600 miles/hour.

Professor Erich von Holst began experimenting with the rotor blades placed in different locations on the *Triebflügel's* fuselage. The first thought was to mount the rotor blades at the nose of the fuselage...as had been the thinking with the double contra rotating propeller blades. Digital image by *Gareth Hector.*

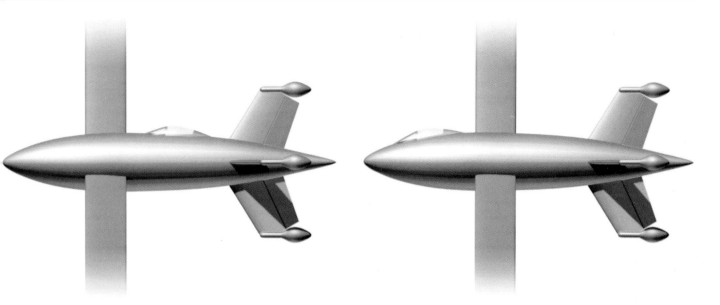

Continuing research regarding the optimum placement of the *Triebflügel's* rotors led *Professor von Holst* to move the rotors, aft as well as the pilot's cockpit. However, it was believed that the pilot would have a very difficult time situated aft the spinning rotors, and this concept was abandoned in favor of putting the cockpit forward the spinning rotors. Digital image by *Gareth Hector*.

This is the final design concept for *Professor von Holst's Triebflügel.* Aerodynamicist *Hans Multhopp* had considerable input into this final configuration. Digital image by *Gareth Hector*.

In late 1944, *Roy Marquardt*, an aviation engineer formerly employed by *Northrop Aviation* on the *XB-35* "flying wing," received a contract from the U.S. Navy Bureau of Standards to develop a 20 inch diameter simple ramjet engine. In the summer of 1946, the Navy had fitted two 20 inch diameter *Marquardt Aircraft* ramjets on the wing tips of a *Grumman F7F "Tigercat."* Each 20 inch ramjet engine weighed 100 pounds. About the same time the U.S. Air Force fitted two *Marquardt Aircraft* 20 inch diameter ramjet engine on a *North American P-51*. During operation they increased the speed of the "*Mustang*" by 40 miles/hour. The Navy continued its testing with different airplanes in 1947 and fitted 20 inch *Marquardt* ramjet to a *Bell Aircraft P-83* and an *North American F-82 "Twin Mustang."* Meanwhile, the U.S. Air Force was seeking bigger ramjets. In 1948 *Marquardt Aircraft* delivered a 30 inch diameter ramjet which weighed 300 pounds and developed approximately 4,000 pounds of thrust each. They were installed on the wing tips of a new *Lockheed P-80A "Shooting Star"* serial number 44-85214. During one test flight which left Burbank Air Terminal, *Lockheed's* home airfield on June 17th 1948, the *P-80A* accelerated to 400 miles/hour. *Lockheed* test pilot *Herman R. "Fish" Salmon*, idled the aircraft's turbojet engine and the twin *Marquardt Aircraft* ramjet engines were ignited pushing the fighter aircraft to 600 miles/hour the first flight of an aircraft powered solely by ramjet engines. During this particular test flight, local police received numerous complaints over the noise created by the operating ramjets. Eventually about 100 test flights were made, mainly out of the U.S. Air Force Muroc Flight Test Facilities. The Air Force's ramjet program

Professor Erich von Holst. **Photographed in the mid-1950s.**

was discontinued when it became apparent that the ramjet consumed fuel at a much too rapid rate to make it a practical means of aircraft propulsion although *Marquardt Aircraft* could produce ramjets for about $1,000 per engine or about .50 cents per ramjet horsepower.

Ramjet Engine Mechanics
Structure, Overall

In simplest terms, a ramjet engine is a long tube with a torpedo-shaped object at the front opening, which is called an inlet. Air rushing through the inlet is compressed between the sides of the tube and the torpedo-shaped body, whose nose is pointed at the front. Its wider part, which sits inside the engine, slows incoming air down to subsonic speeds. Fuel is then sprayed into the air stream, and the mixture is ignited as it flows past a flame inside the combustion chamber. The resulting gases are then ejected out of a nozzle in the back, yielding plenty of thrust.

Air Intake

The classic subsonic ramjet engine has an annular inlet which directly faces the air stream. The inlet is about half the area of the combustion chamber and smoothly shaped to reduce drag on the air flowing around it.

Air enters the inlet at about the speed of the flying machine. Because the inlet is narrowed at the front the air splits with some of it going around the outside and some of it going in. The air that goes outside flows smoothly around the cowling-covered tube causing very little drag. The air that goes inside (intake) finds a widened area called the diffuser and spreads out. When it does it slows down and increases in pressure (compresses).

Diffuser

Usually right behind the inlet is the subsonic diffuser which slows the air and thereby compresses it. On some ramjets the air enters the inlet near the front of the engine and travels along a duct before entering the diffuser. On some ramjets the expansion may be accompanied with a dramatic change in shape.

Combustion

The combustion chamber is an open area inside the ramjet where the fuel/air mixture can burn completely before being exhausted out the nozzle. The combustion chamber includes the flame holder. The combustion chamber is typically the wildest part of the ramjet engine in terms of cross sectional area.

One of the earlist pen and ink illustrations of *Erich von Holst's* **three rotor vertical take off and landing flying machines powered by a piston engine installed in the forward fuselage. Notice the absence of any ramjet engines on the wing tips.**

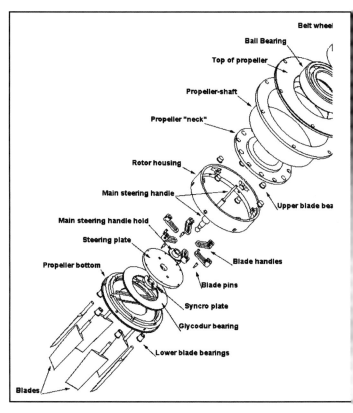

About this time *Professor von Holst* was thinking a suitable power plant would be one which used the *Voith-Schneider* propeller (1933); its controllable blade angles had already been applied to ship propulsion. To *Professor von Holst* the *Voith-Schneider* controllable blade propeller was the successful development of the functions of the outer part of a bird's wing. It is not quite clear how *von Holst* would have applied the *Voith-Schneider* blade propeller to power the *Triebflügel*.

After the air enters the inlet, fuel is added via injectors to provide a combustible mixture. This fuel/air mixture is then ignited which then raises its temperature and increasing the volume so that it exits out the nozzle at a much higher velocity than it came in with. The air entering the combustion chamber is moving at about 200 miles/hour. To ensure that the flame doesn't blow out under these conditions a flame holder is placed in the air stream. There are several variations on flame holders and their location inside the duct for injecting fuel.

Fuel Injector

The fuel injector adds fuel to the air to provide a combustible fuel/air mixture. This is best accomplished when the fuel is vaporized or at least atomized into very small droplets and thoroughly mixed with air. The fuel is typically sprayed in from nozzles, often through a spray bar.

In some cases it is desirable to have an uneven mixture of fuel and air. If the engine is running at its highest design temperature (stoichiometric) then a complete mix is best. But if it is running lean or rich in order to effect thrust, range, or temperature, the a flame may be difficult to maintain. In this case a turbulent, uneven mixture will have some regions which are rich, some that are lean, and more importantly, some that are stoichiometric. It is these regions that will sustain the flame and ignite the other regions.

Flame Holder

The purpose of the flame holder is to provide an environment where the fuel/air mixture will burn without blowing out. The fuel/air mixture is typically traveling through the engine at about 200 miles/hour or 300 feet/second making it difficult to keep the flame lit. The classical flame holder is the gutter type. This is just a "V" or "U" shaped piece of metal with the open side facing downward. Behind, and inside, the gutter the fuel/air mixture is highly turbulent with small pockets of slow moving eddies. It is in these eddies that the flame is actually held, spreading to the rest of the fuel/air mixture. Designing a low-drag, efficient flame holder is an art.

Igniter

The function of the igniter is to ignite the fuel/air mixture, that is, to get the flame started. The classical way is a short burning flare in the flame holder. It can also be done with a spark, addition of special chemicals such as phosphorous, or a small torch. It is also possible to add a continuous spark so that the flame can be restarted if it blown out.

A pen and ink drawing of *Hero's* aeolipile, which is reputed to be the first apparatus converting steam pressure into mechanical power. It probably was the earliest demonstration of the principle of jet reaction. *Hans Multhopp* believed that *Professor von Holst's Triebflügeljäger* might also benefit from the principle of jet reaction by placing small jet units on the rotor tips.

Suggested in 1903 — a helicopter using *Avery's* rotary engine with its steam jet reaction. This was the same principle which *Hans Multhopp* of *Focke-Wulf* was thinking for the *Triebflügel*. The challenge would be to miniaturize a jet-type engine for placement on the rotor tips.

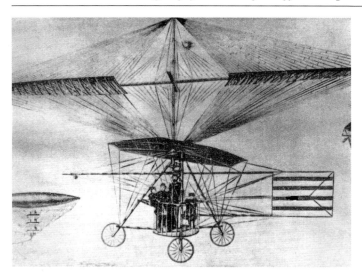

A pen and ink drawing featuring rocketeer *Hermann Ganswindt's* "flying machine" with its rocket propelled helicopter-like lifting arrangement. From the turn of the century, about 1900.

Exhaust

The purpose of the exhaust nozzle is to maintain pressure in the combustion chamber by restricting the outflow of the gases. The classical nozzle is circular but may be more complex. It may be oval, square, rectangular, or even more complex. Highly complex nozzle exits may have moveable surfaces to improve efficiency, but these nozzles are very expensive and only rarely used.

As the fuel/air mixture burns to become exhaust products it heats up and accelerates toward the exhaust nozzle. The exhaust nozzle throat provides a constriction to help maintain pressure inside the combustion chamber. The hot exhaust products flow/pass throat and expand in the nozzle exit to approximately atmospheric pressure.

Thrust

The ramjet's net thrust is the difference between two opposing forces, drag, and thrust. Drag comes from the intake air being slowed down (for compression), flow around internal components like the flame holder, and flow along the skin (internal and external). Thrust comes from heating the air to high temperatures and accelerating it. To maximize the net thrust you need to minimize the drag and maximize the thrust. This means designing an inlet that flows the right amount of air smoothly, finding the lowest drag flame holder that works, keeping the air smooth through the engine, and designing a nozzle of the correct size. To maximize thrust you need the highest temperature one can achieve. This comes when the fuel to air ratio is stoichiometric, that is, every atom of oxygen in the air has the correct number of fuel atoms to go with it.

A Variety of Ramjet Powered Aircraft At War's End

By war's end a growing number of German aircraft companies had proposed ramjet powered flying machines. In addition to *Focke-*

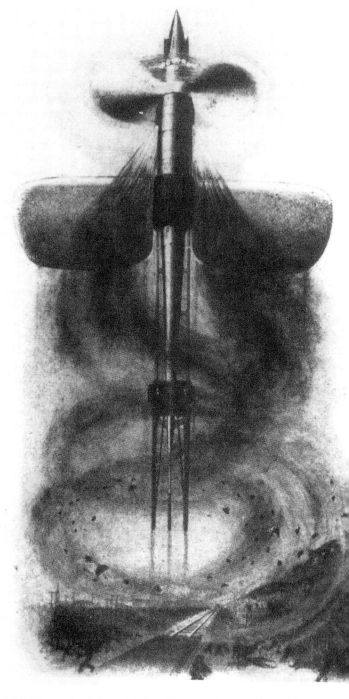

This illustration is by *Dipl.-Ing. Hermann Langner* from his two-part article *Die geschichte einer Mars-expedition,* or The Story of a Mars Expedition. The article appeared in October 1922. The *Langner* rocket ship is shaped like a spindle with an extremely long tail, tapering to a needle-like point. Near the nose is a giant propeller, and amidships are a pair of short, broad wings. The *Langner* rocket ship takes off vertically, aided by the propeller, which would also have been used for maneuvering in the atmosphere of Mars.

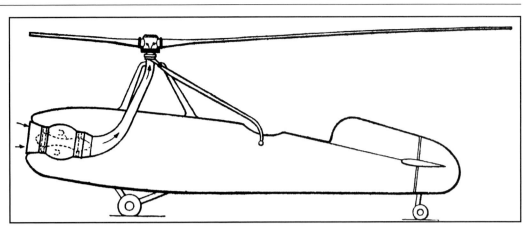

The *Weir* helicopter with rotors driven by jet reaction and exiting at the rotor blade tips. This was the principle suggested to *Professor Erich von Holst* for his vertical take off and landing *Triebflügel* by *Focke-Wulf* aerodynamicist *Hans Multhopp*. The turbine compressor unit was to have been mounted in the nose of the fuselage.

Wulf Flugzeugbau's Triebflügeljäger, other ramjet powered projects included:

• *von Braun A-9* piloted rocket with a *Sänger* supersonic ramjet;
• *Dornier Do 17* with *Pabst* ramjets;
• *Dornier Do 217* with *Sänger* ramjets;
• *Sänger* long-range subsonic fighter;
• *Sänger* long-range supersonic fighter;
• *Sänger* ramjet fighter variants;
• *Skoda Sk Projekt 14* fighter with *Sänger* ramjets;
• *Messerschmitt Me 262* with *Sänger* ramjets;
• *Messerschmitt Me Projekt 1101L* with *Sänger* ramjets;
• *Heinkel He Projekt 1080*;
• *Focke-Wulf Fw Projekt 283* bomber with *Pabst* ramjets:
• *Focke-Wulf Fw Super Lórin Projekt* with *Pabst* ramjets;
• *Messerschmitt* combination rocket/ramjet transport;
• *Lippisch "Triebflügel"* ramjet *projekts*;
• *Lippisch Li Delta V*;
• *Lippisch Li Projekt 12*;
• *Lippisch Li Projekt 13*;
• *Lippisch Li Projekt 13b*;
• *Lippisch Li "Uberschallflügel;"*
• *Lippisch Li* ramjet powered fighter *projekts*;
• *Lippisch Li DM-1* and variants;
• *Blohm & Voss* manually controlled rocket *projekts* and variants;

Ramjet Research and Development Activities

Otto Pabst had been investigating ways to reduce the ramjet's overall diameter and length yet increase the thrust of conventional *Lórin* ramjet engines following the success' of rocketeer *Dr.- Ing. Eugen Sänger. Sänger* at that time (1943) was employed by *Professor Walter Georgii*, head (*DFS*). On the other hand, *Otto Pabst* was in *Focke-Wulf Flugzeugbau*-Bad Eilson's Department of Gas Dynamics. He was highly interested in the development of special burners and in the methods of air compression as it entered the ramjet's air intake. It appears that *Pabst* and the *Luftfahrtforschungsanstalt Hermann Göring (LFA)* or The *Hermann Göring* Aeronautical Re-

search Establishment's wind tunnel expert *Zobel* had designed a ramjet engine which had a reduce drag coefficient due to its clean aerodynamic shape. They were anticipating that their ramjet would work efficiently up to altitudes of 59,000 feet [18,000 meters]. The first operating prototype is believed to have been ready in August 1944. It has been suggested that actual testing did not begin for

Dipl.-Ing. Baron Friedrich von Doblhoff.

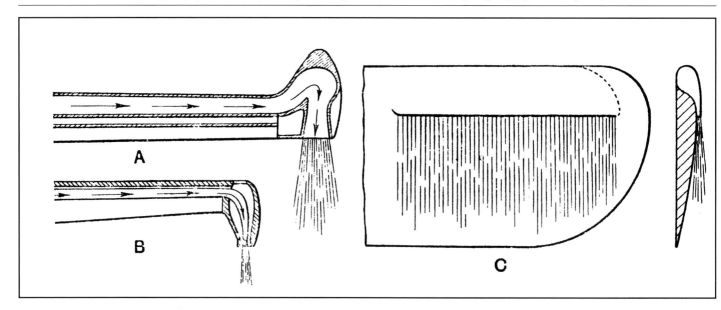

In the *Weir* helicopter, the turbine-compressor unit mounted in the nose of the fuselage forces its exhaust out to the rotor tips via a hollow hub of the rotor, where it is distributed to the hollow blades for discharge from reaction nozzles at the rotor tips. Suggestions for various reaction nozzles included: A - this illustration features an "elbow" duct with the reaction nozzle terminating in an elongated orifice to conform to the blade profile. On the leading edge, opposite the nozzle, is a weight for the mass balancing of the blade. With the weight well forward it serves to prevent blade flutter. B - an alternative reaction nozzle is featured which the balancing weight is embodied in the leading edge, and the nozzle is an ejector type comprising a chamber with an air inlet at the front edge and a venturi nozzle at the rear. Into the throat of this nozzle the working fluid is discharged by an ejector nozzle, and thus the mass flow is augmented by entrained air. C - the third type of reaction nozzle constitutes a combination of the hollow blade spar forming the working fluid duct and discharges through a slot-like orifice over the upper or low pressure surface of the blade at about the region of maximum thickness. The thrust is applied close to the mass axis of the blade, and the jet is discharge where the external air stream has its maximum local velocity and tendency to separate from the blade profile.

The *Wiener-Neustädt Flugzeugwerke Wn 342-4 V4* two-seat helicopter with its rotor tip reaction nozzles/combustion chamber driving the rotor blades, thus providing lift. Shown also in the photo is its designer, *Dipl.-Ing. Baron Friedrich von Doblhoff.*

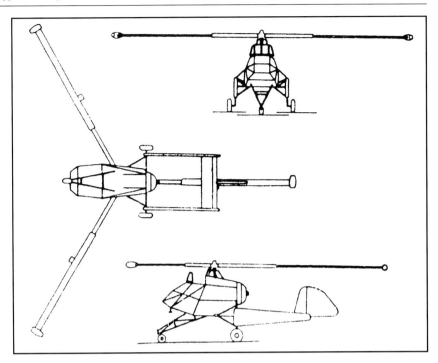

A poor quality three-view pen and ink illustration of the *Wiener-Neustadt Flugzeugwerke Wn 342-4*. Rotor blade diameter was 32 feet 9 1/2 inches [10.0 meters].

several month later due to Allied air raids and delays in obtaining the promised correct fuel for the ramjet engine.

Dr.-Ing. Theodor Zobel had developed his interferometer method for the study of high-speed air flow in *LFA's*-Braunschweig wind tunnel A.2 with its 9.2 foot diameter working section, in which Mach numbers of 0.8 to 0.95 were obtaining at maximum power of 12,000 kilowatts. It was in the A.2 wind tunnel which *Zobel*, working in co-operation with the *Argus* and *Fieseler* companies that the pulse jet power plant for the *Fieseler Fi 103 V-1* "flying bomb" had been developed.

It became increasing evident that *Otto Pabst's* small diameter short length ramjet designs could be put on the tips of each of the three rotor blades, or wings, of *Professor von Holst's Projekt Triebflügeljäger*, resulting in a combination that avoided the problems encountered earlier: the complex reduction gear, excessive weight, and an inability to supply power to handle the contra-rotating propellers. The ramjets appeared to be a "natural" for the *Focke-Wulf Flugzeugabu* VTOL, and by 1944 the need for an aircraft with VTOL capabilities was becoming oblivious. With the *Luftwaffe's* loss of air superiority, operations from conventional bases, which

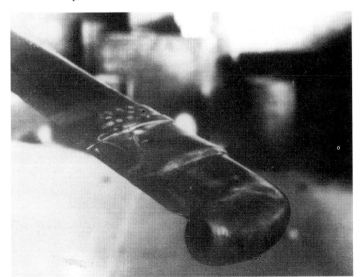

Wiener-Neustadt Wn 342-4, showing the reaction jet nozzle on its rotor tip. Fuel consumption is reported to be substantial...31 gallons per hour [140 liter/hour].

Dipl.-Ing. Baron Friedrich von Doblhoff is seen in the pilot seat of his two-seat *Wiener- Neustädt Flugzeugwere Wn 342-4* rotor jet nozzle propelled helicopters, as seen from its port side. Fuel consumption was extreme for this experimental helicopter. Photo taken in war-time Germany.

A rear port side view of the rotor bladed jet nozzle propelled *Wiener-Neustädter Flugzeugwerke Wn 342-4*. Photo taken in war-time Germany. The flying machine's flight-ready weight was 1,401 pounds [640 kilograms].

were under attack around the clock, became very expensive and exhausting in terms of military manpower and material. A ramjet-powered *Triebflügeljäger*, it was believed, could be placed throughout Germany, in the valleys, mountains, forests, even cities, and would not have to rely on airstrips in order to carry out its mission as a bomber interceptor.

A tail-sitting VTOL aircraft such as *Focke-Wulf's Triebflügeljäger* which derived its lift from three wings (one project was reported to be a two wing version while another project is mentioned having four wings) which rotated around the fuselage immediately aft of its cockpit. The three rotor version, for example, would have utilized three un-tapered variable incidence wings, fixed to a rotary collar located about one-third of the way down the fuselage from the nose. They were to rotate around the fuselage transmitting little or no torque to the fuselage. Under the power of tip-

A port side view of a rotor jet nozzle propelled *Wiener-Neustädter Flugzeugwerke Wn 342-4*. It also achieved forward motion thanks to its single *BMW Brano Sh 14A* 140 horsepower piston engine. Its maximum forward speed was about 31 miles/hour [50 kilometers/hour]. Photo taken in the United States post war.

The *"Roton"* had four rotor blades, each with a small 85 percent pure hydrogen peroxide (H_2O_2) powered rocket on the tip. Each rocket provided up to 350 pounds thrust to power the rotor as seen in this photo.

mounted *Pabst* ramjets producing about 1,852 pounds [840 kilograms] of thrust each, the wings would be brought up to operating speed by probably three jettisonable *HWK* rocket engines. It is speculated that each ramjet engine might have a built in *HWK* bi-fuel liquid or solid fuel starter engine. In flight, the wings would be rotated around their individual axis until they became conventional airfoils as the aircraft itself rotated until its axis was horizontal rather than vertical. During vertical lift-off, level flight, and vertical landing the pilot's seat would have remained fixed in one position. A fixed position seat wold have meant that the pilot was in a reclined position during the most difficult moments of the *Triebflügeljäger's* flight...take off and landing.

No development work is known to have been carried out on *Focke-Wulf's Triebflügeljäger* concept, and the viability of the design is a matter of speculation. However, three tail-sitting VTOL aircraft were built postwar, two in the United States and one in France. In addition, we have the ramjet powered *Hiller HOE-1* helicopter of the late 1940s. The two American tail-sitting designs, from *Lockheed* (*XFV-1*) and *Convair* (*XFY-1*), were somewhat more conventional, in that they used fixed wings and contra-rotating propellers in the nose. The French version from *SNECMA* known as the *C.450.01* *"Coléoptère"* was powered by a tail-mounted turbojet engine and had an annular wing. Control was achieved through four swivelling fins. All three aircraft flew but each of the projects were eventually canceled. The objective has since been achieved by vectoring the thrust of a turbojet engine in a modified conventional airframe.

The so-called "Rotary Rocket" of 1999, or "Roton" by the *Rotary Rocket Company*, Redwood Shores, California. Known as the *Atmospheric Test Vehicle (ATV)*, it was an idea for a small reusable satellite-launch vehicle. The *ATV* would take off vertically, powered by a spinning engine that contained 72 small rocket engines arranged around its base in a circle. After delivering a payload to low Earth orbit, the craft would turn around and unfold its helicopter blades. It would reenter Earth's atmosphere, base first. The project flew twice...July 23rd 1999, at a height of about ten feet off the ground. The second test flight on September 16th 1999 saw the *"Roton"* hover 10 to 15 off the ground for as long as 2 1/2 minutes. Investors failed to show interest in the concept, and the *Rotary Rocket Company* abandoned their *"Roton"* in early 2001 as being "too technically ambitious."

A rear port side view of the experimental rotor jet nozzle propelled *Wiener-Neustädter Flugzeugwerke Wn 342-4* shown after its arrival in the United States post war. *von Doblhoff* was invited to the United States post war. He worked for *McDonnell Aviation* doing helicopter research and development.

One of the four hydrogen peroxide 350 pound thrust rocket motors on the tip of the *"Roton's"* rotor blade to keep the rotor spinning during the final flare to touchdown. The four rocket motors consumed about 1,000 pounds of 85 percent pure hydrogen peroxide a minute. Carrying 4,300 pounds of hydrogen peroxide, flights were limited to two to three minutes. Eighty-five percent pure hydrogen peroxide cost over $5 dollars a pound in 1999.

Specifications of the *Otto Pabst Lórin* Ramjet Engine
• **Diameter, Overall:** 2 feet 3 inches [68.58 centimeters];
• **Length, Overall:** Not greater of two times average (perhaps 2 1/2 times) its own diameter;
• **Thrust:** 1,852 pounds [840 kilograms];
• **Wind Tunnel Test Operating Speed:** up to Mach 0.9 [667 miles/hour]
• **Fuel:** nonstrategic and low quality fuels such as powdered coal, crude oil, gaseous liquid which could be injected as a vapor into the operating ramjet engine (tested using propane);
• **Horsepower:** 3,400 estimated horsepower each;
• **Minimum Operating Speed:** 186 miles/hour [300 kilometers/hour];
• **Ramjet Starter Engine:** 660 pound thrust *HWK* rocket motor (unknown if bi-fuel liquid or solid fuel);
• **Ramjet Starter Engine Location:** Unknown if located inside each ramjet engine, on the trailing edge of each rotor, or inside the aircraft fuselage centerline;
• **Rotor Location on Fuselage:** 3 rotors were located approximately 1/2 the length of the fuselage back from the aircraft's nose;

Hans Multhopp in the United States post war. He is seen holding a scale model of a lifting body designed by his employer *Martin Aviation*, Baltimore, Maryland.

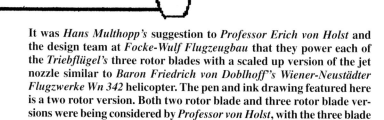

It was *Hans Multhopp's* suggestion to *Professor Erich von Holst* and the design team at *Focke-Wulf Flugzeugbau* that they power each of the *Triebflügel's* three rotor blades with a scaled up version of the jet nozzle similar to *Baron Friedrich von Doblhoff's Wiener-Neustädter Flugzwerke Wn 342* helicopter. The pen and ink drawing featured here is a two rotor version. Both two rotor blade and three rotor blade versions were being considered by *Professor von Holst*, with the three blade rotor becoming the preferred version.

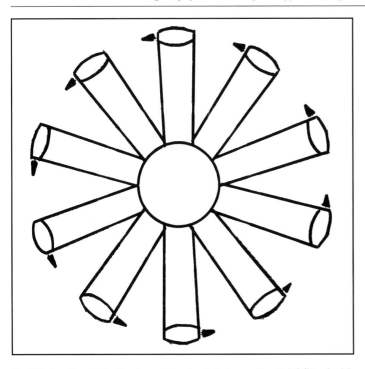

An illustration featuring how the rotor blades on the *Triebflügel* with their blade tip jet nozzles would appear during vertical lift off or landing, as seen from above.

• **Rotor Speed during lift-off and landing:** approximately 220 revolutions per minute;

• **Rotor Speed during forward flight:** rotors would be in a fixed position and acting as an airfoil providing lift;

• **Maximum Operating Altitude:** 59,000 feet [18,000 meters];

• **Number Built:** None known to completed by war's end, although *RLM's Oberst Siegfried Knemeyer* ordered that four airworthy *Lórin* ramjets were to have been ready by August 1945. It is not known for sure if in fact they were;

The *Projekt Triebflügeljäger* interceptor was to stand vertically on the ground, supported by four tail fins, each of which had a

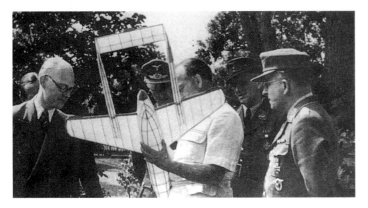

One of the leading members of the *Triebflügel's* fuselage design team was *Focke-Wulf's* designer *Ludwig Mittelhüber*. He is on the far left. *Kurt Tank*, *Focke-Wulf 's* managing director, is seen holding a wood stick and paper model of their proposed turbojet powered *"Flitzer"* fighter.

A group of aircraft designers from *Focke-Wulf Flugzeugbau*: 1 - *Käther*; 2 - *Stampo*...holding his wood stick and paper scale model of *Hans Multhopp's Ta 183* proposed turbojet powered fighter; 3 - *Ludwig Mittelhüber*...member of the *Triebflügel* design team; 4 - *Otto Pabst* - gas dynamicist and inventor of the miniature ramjet engine of the type to power the *Triebflügel*; and 5 - *Naumann*.

The Frenchman *René Lorin*...inventor of the ramjet engine.

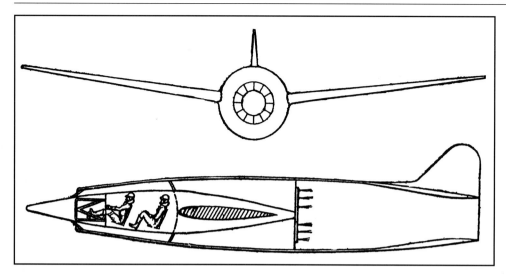

The French aero engineer *René Leduc* is mainly associated with projects for compressorless jet-propelled aircraft, so-called "athodyds," in which the main fuselage structure forms a continuous air duct from nose to tail. In 1938, at the Paris Salon de I'Aviation, a model of a *Leduc* aircraft was exhibited. A notice suggested that this "machine of the future" would have an output of 14,000 horsepower, a speed of 621 miles/hour [1,000 kilometers/hour], and ceiling of [30 kilometers]. Nothing further was heard of this design. In the interval, *Eugen Sänger* and *Otto Pabst* made considerable progress in *Nazi* Germany.

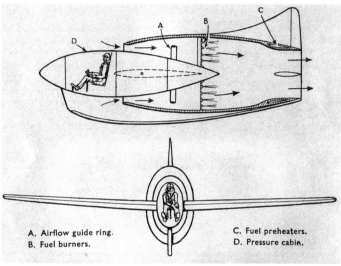

A. Airflow guide ring.
B. Fuel burners.
C. Fuel preheaters.
D. Pressure cabin.

René Leduc, in 1933, proposed this ramjet propelled flying machine. The fuel was preheated and delivered by a series of nozzles distributed across the main airduct.

smaller outrigger wheel at its tip. The main landing load was to be taken on a single main wheel at the base of the fuselage. During flight all wheels were to be enclosed by streamlined, tulip-shaped pods. *Projekt Triebflügeljäger's* pilot was to be accommodated conventionally in a nose cockpit. Armament which consisted of two 30 mm *MK 103* cannon with 100 rounds of ammunition each, plus two 20 mm *MG 151* cannon with 250 rounds each was to be placed in the nose.

Control of the aircraft was to be accomplished by means of control surfaces at the trailing edges of the tail fins. For flying in a horizontal position the tail would be depressed slightly to direct part of the thrust force into a lift force. In addition, the blades or wings, could be rotated. During lift-off they would provide the lift needed to get the aircraft off the ground. As it gained altitude, each of the blades would be rotated about its axis and take on more of the function of a fixed wing, while the ramjets at each blade tip would provide the thrust for forward flight. Output of each ramjet was estimated at 3,400 horsepower, more than enough power to lift the 11,385 pound [51,642 kilogram] interceptor and propel forward

The brilliant, clever, and highly secretive head of the *Reichluftfahrtministerim (RLM)*, *Oberst Siegfried Knemeyer*. He was very interested/supportive in *Professor Erich von Holst's* vertical take off and landing *Triebflügel* flying machine.

Professor Walter Georgii, chairman of the four-man *Reich* Aero Research Commission...a very powerful group which assigned priorities to aircraft research and development. *Georgii*, along with *Siegfried Knemeyer*, was highly supportive of *Professor Erich von Holst's Triebflügel* idea. With *Knemeyer* and *Georgii's* support, the *Triebflügel* had top priority within German aircraft circles.

Professor Kurt Tank, head of the *Focke-Wulf Flugzeugbau* aircraft organization. It is unclear what Tank thought of the *Triebflügel* concept, however, several very powerful individuals within *Nazi* Germany's aviation research and development community were highly supportive. *Tank* probably went along with those individuals, such as *Oberst Knemeyer* and *Professor Georgii,* and supported its research and development.

Dr.-Ing. Eugen Sänger of *DFS. Sänger,* the rocketeer, is probably best known for his proposed suborbital, manned, reusable, horizontally launched bi-fuel liquid rocket flying machine known as one of *Oberst Siegfriend Knemeyer's* "*Amerika Bomber*" projects.

it through the air on the order of 575 miles/hour [925 kilometers/hour] at sea level. Forward speed would have decreased considerably at higher altitudes. For example, it was estimated its forward speed at 45,960 feet altitude would have dropped to 419 miles/hour [675 kilometers/hour] a loss of 156 miles/hour. Regardless of its projected loss of performance at high altitudes, a two other *Triebflügeljäger* versions have been mentioned in reports: a twin-bladed and a four-bladed versions were considered. No performance data is available on the two rotor version. *Focke-Wulf Flugzeugbau* engineers believed that a four-blade arrangement would create excessive drag and, consequently, unsatisfactory performance. Nevertheless, no one had the slightest idea if the concept would even lift itself off the ground, transition to forward flight, land, and if so how well. Yet, several German helicopter designers were testing helicopters powered by small jet thrusters located in their rotor tips

such as *Baron Friedrich von Doblhoff's Wiener Neustädt Flugzeugwerke Wn 342-4.* Compressed air mixed with vaporized fuel was channeled through the rotors to wing tip combustion chambers where the mixture was ignited producing a jet- like thrust.

The *Wiener-Neustädter Flugzeugwerke Wn 342-4* was a wing tip jet powered helicopter. The idea had been developed by the Austrian *Baron Friedrich von Doblhoff* with the collaboration of *Dipl.-Ing. Theodor Lauft* and *Dipl.-Ing. Stephan* at *Wiener-Neustädter*-Vienna. Although it never saw service the prototype, which was begun in October 1942, featured a 60 horsepower *Walter Micron* piston engine. This motor supplied power to run an *Argus* supercharger. The compressed high-speed air leaving the supercharger was then pre-heated and vaporized fuel was injected. The mixture was then routed out the rotor tips via a pipe running the length of the rotor. At the rotor tip the mixture was introduced into the combustion chamber where it was ignited. Testing of the ramjet-like powered helicopter was very successful from the very beginning. The last prototype had been flown at speeds up to 28 miles/hour [45 kilometers/hour]. At war's end the United States gathered up each of the four *Wiener-Naustädter Wn 342-4* prototypes, research material, and shipped it all to America.

Focke-Achgelis also produced several VTOL design prior to war's end. One, the *Focke- Achgelis Fa 269* was known as a "convertiplane" would have taken off and landed vertically. For forward flight the rotors would have turned 90 degrees to bring them in a position of pusher propellers. This idea was not successfully accomplished until the late 1980's when *Boeing-Vertol* introduced their *Vertol V-22 "Osprey."*

Specifications: *Focke-Wulf Triebflügeljäger*
• **Engine**: 2, 3, and 4 rotor version *Lórin* ramjet engine version from *Focke-Wulf Flugzeugbau* with the 3 ramjet version thought to

Professor Walter Georgii was also head of the *Deutsches Forschungsinstitut für Segelflug*- Ainring (*DFS*), and was a tireless promoter of *Dr.-Ing. Eugen Sänger* and his research and development on the *Lórin* duct...the same type of engine *Hans Multhopp* suggested to *Professor von Holst* to power his *Triebflügel* vertical takeoff and landing flying machine.

The *Dornier Do 217-E-2 (RE+CD)* of *DFS*-Ainring carrying one of *Eugen Sänger's* experimental *Lórin*-type ramjet engines and seen from its rear starboard side. Notice that the ramjet engine carries a number "6" on its side. It is not entirely clear to this author what that number means.

be the final design version, each having 1,852 pounds [840 kilograms] of thrust, brought up to ramjet operating speed by 2 or 3 solid fuel rockets installed in the ramjet and having 661 pounds [300 kilograms] of thrust.

• **Wingspan**: 37 feet 7 inches [ll.5 meters];

• **Wing Area**: 177.6 square feet [16.5 square meters];

• **Wing (rotor) Length**; 30 feet 2 inches [9.2 meters];

• **Wing Height**: not available;

• **Height:** 30 feet [9.15 meters] however, *Otto Pabst* writing in book postwar stated that the height of his *Focke-Wulf Fw Triebflügeljäger* would have been 30 1/2 feet [9.35 meters];

• **Weight**, **Empty**; 7,055 pounds [3,200 kilograms];

• **Weight**, **Takeoff**: 11,355 pounds [5,150 kilograms];

• **Number of Crew**: 1;

• **Take-Off:** rotors were positioned by the pilot with a slope of +3 degrees. Forward flight was achieved by gradually tilting the rotors;

• **Rate of Climb at Sea Level:** 410 feet/second [125 meters/second];

• **Rate of Climb at 22,965 Feet [7,000 meters]:** 164 feet/second [50 meters/second];

• **Climb to 3,280 feet [1,000 meters]:** 8.2 seconds;

• **Climb to 49,210 feet [15,000 meters]:** 11.5 minutes;

• **Speed, Cruise**; not available;

• **Speed, Design Maximum at Sea Level**; 575 miles/hour [925 kilometers/hour];

• **Speed, Design Maximum at 45,930 Feet:** 419 miles/hour [675 kilometers/hour];

• **Landing:** transition to vertical flight from forward flight

• **Radius of Operation**: 404 miles [650 kilometers] at sea level and 1,491 miles [2,410 kilometers] at 45,934 feet [14,000 meters];

• **Service Ceiling**: 45,934 feet [14,000 meters];

• **Armament**: 2x30 mm *MK 103* cannon with 100 rounds each and 2x20 mm *MK 151* cannon with 250 rounds each. The possibility

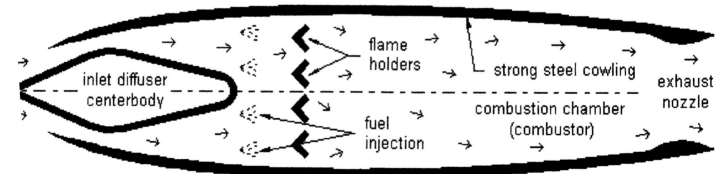

In essence, the ramjet is nothing more than an open cylinder with an internal fuel spray. The device requires a minimum air flow velocity of about 300 miles/hour to become effective, and this is usually obtained by means of a solid or liquid propellant booster. At about 300 miles/hour the air-fuel mixture is ignited and burns contiguously as long as the air stream is passing through the duct at that minimum rate of flow the gases emerging from the exit with a greater energy than that of the air-stream entering the intake. The differential of exit and entry, resulting in a higher exit velocity, imparts a very strong thrust to the duct by the principle of reaction.

An original *Eugen Sänger/Irene Bredt Lórin*-type ramjet engine powered design for a fighter. The fighter was proposed to be equipped with a single 60,000 horsepower *Lórin*-type ramjet engine. The plans for this fighter were turned over to *Skoda-Kauba*-Cakowitz (near Prague) for development. It was given the *RLM* designation of *Skoda-Kauba Sk P.14-01*. Computer generated digital image by *Josha Hildwine* and shown from its nose port side.

A closeup view of the proposed miniature *Lórin*-type ramjet engine prone pilot fighter design from *Eugan Sänger/Irene Bredt* and to be developed by *Skoda-Kauba*-Cakowitz as the *Sk P.14-02*. This digital image, by *Marek Rys*, is seen from its nose port side.

has been suggested in some reports that the "*Triebflügeljäger*" might have been equipped with state-of-the-art high firepower 4x20mm *MK 213* cannon;
• **Camouflage:**
• *RLM 74* - Dark Gray upper surfaces;
• *RLM 75* - Medium Gray upper surfaces;
• *RLM 76* - Light Blue Gray under surfaces;
• **Unit Markings:** None since no "*Triebflügeljäger*" was constructed;
• **Landing Speed**: not available;

• **Supporting Carriage:** entire flying machine rested vertically on the ground supported by its 4 auxiliary empennage wheels which consisted of a extendable (retractable) gear with a wheel at its end plus a center wheel, located at the base of the fuselage. During flight the wheels were covered;
• 4 empennage wheels with a diameter of 380x150 each;
• 1 main wheel with a diameter of 780x260;
• **Rate of Climb**: 24,608 feet/minute [7,500 meters/minute];
• **Bomb Load**; not available;

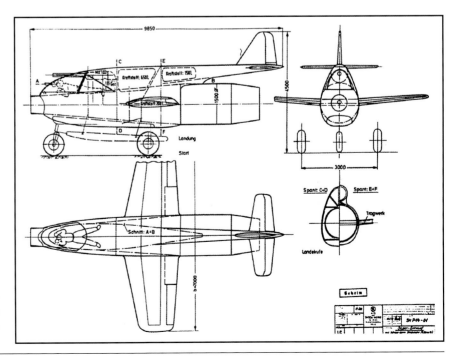

A drawing of the proposed *Eugen Sänger/Irene Bredt, Lórin*-type ramjet engine powered *Skoda-Kauba Sk P.14-01*.

The proposed *Lórin*-type ramjet engine powered miniature fighter design by *Eugen Sänger/Irene Bredt* and to be developed by *Skoda-Kauba* as the *Sk P.14-02*. Digital image by *Marek Rys* and shown from its rear starboard side.

• **Flight Duration**; 42 minutes at sea level and between 2 hours 30 minutes and 3 hours and 24 minutes at 45,934 feet [14,000 meters] altitude;

Post War Vertical Take-Off and Landing (VTOL) Prototypes and Test Results

In post WWII several countries experimented with vertical take-off and landing vehicles or VTOL. The *Convair XFY-1* and *Lockheed XFV-1* were radical experiment single seat shipboard fighters evolved as part of a program to examine the feasibility of operational VTOL aircraft from small platforms on ships as an alternative to carrier launching. The principal mission of the fighter was

Dr.-Ing. Otto E. Pabst of *Focke-Wulf Flugzeugbau*. His goal was to take the research and development of the *Lórin*-type ramjet done by *Eugen Sänger* and develop a shorten version...one which would fit nicely, for example, on *von Holst's* vertical take off and landing flying machine.

A *Messerschmitt Me 262A-1a* shown here with proposed twin *Eugen Sänger Lórin*-type ramjet engines in addition to its twin *Junkers Jumo 004B* turbojet engines. This digital image by *Mario Merino* features the *Messerschmitt Me 262A-1a* as seen from its rear port side.

the protection of convoys from aerial attack, and both the *XFY-1* and *XFV-1* were designed to take off vertically from an erect position. They were designed to meet the requirements of a specification prepared by the U.S. Navy in 1950. Prototypes of the two designs were ordered in March 1951. The first to fly was the *Lockheed XFV-1* which began normal horizontal trials in March 1954 with the aid of a temporary fixed undercarriage. A 5,850 horsepower *Allison T40-A-6* turboprop was installed for initial tests pending the delivery of the *Allison YT40-A-14*. This engine was expected to provide 7,100 horsepower for take-off driving broad bladed contra-rotating propellers. Full performance trials with the *XFV-1* were not completed, however, it was calculated as follows:

• test weight of 15,002 pounds;
• maximum speed of 580 miles per hour;
• a rate of climb to 20,000 feet of just under 3 minutes;

Lórin duct research was attempted for a time by America's NACA. This photo is from their research from around February 1941 and it featured an 11 inch diameter ramjet. It is reported that the duct was wind tunnel tested up to speeds of .075 Mach [556 miles/hour] and registered a thermal efficiency of 9.5 percent, close to the theoretical ideal of the perfect 10.0 percent.

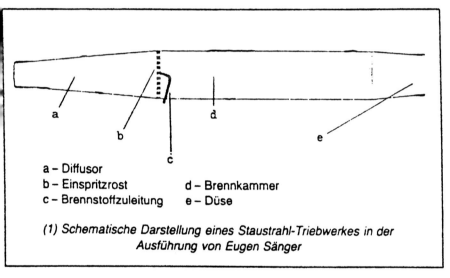

a – Diffusor
b – Einspritzrost   d – Brennkammer
c – Brennstoffzuleitung   e – Düse

(1) Schematische Darstellung eines Staustrahl-Triebwerkes in der Ausführung von Eugen Sänger

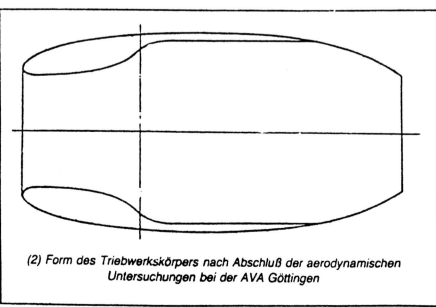

(2) Form des Triebwerkskörpers nach Abschluß der aerodynamischen Untersuchungen bei der AVA Göttingen

This pen and ink drawing illustrates the differences between *Eugen Sänger's* ramjet engines of *DFS* and that of *Otto Pabst* of *Focke-Wulf*. Otto Pabst wanted to miniaturize the ramjet engine to fit on the rotor blades of the *Triebflügel*.

***Professor Dr.-Ing. Albert Betz* (left) with *Dr.-Ing. Theodor von Kármán* post war late May 1945.**

The *Convair XFY-1* had a similar turboprop installed for its first vertical flight on August 1st 1954. Although both the *XFV-1* and *XFY-1* made a number of successful transitions (vertical flight to horizontal flight), they presented severe piloting difficulties, and the program was abandoned by the United States Navy.

The French *SNECMA* experimented with a VTOL aircraft in the late 1950s. Known as the *C.450.01 "Coléoptère"* it was essentially a tail-sitter with an annular wing intended to operate in a similar fashion to the *Lockheed XFV-1* and *Convair's XFY-1*. The *"Coléoptère"* had a take-off weight off 6,615 pounds with an esti-

mated performance of 497 miles per hour. It was powered by an 8,150 static thrust *SNECMA Atar E5V* turbojet. Directional control was attained by means of jet deviation. Directional control during horizontal flight was provided by four swivelling fins. Initial vertical take-offs and landing were made on April 7th 1959. Shortly afterwards the *"Coléoptère's"* test pilot, *Auguste Morel*, was forced to eject from the *C.450.01* at an altitude of less than 230 feet after completing several inclination maneuvers, the aircraft being almost totally destroyed. The result of this crash brought an end to the *"Coléoptère"* program.

The *C.450.01* is interesting, too, for another reason. It was designed by the high-ranking *SS* member *Dipl.-Ing Helmut von Zborowski*. During WWII, *Zborowski* worked at *BMW*- Spandau (later as an assistant to *Dr.-Ing Eugen Sänger* along with assistant *Irene Bredt* on *Sänger's* piloted, horizontally launched sub-orbital rocket bomber project) and was responsible for the development of the bi-fuel liquid rocket engine attached to a standard *BMW 003* turbojet engine. This take-off assist rocket was intended for the *Arado Ar 234B*, *Ar 234C*, and the *Heinkel He 162 "Volksjäger."* The *BMW 003R* combined turbine and rocket engine, as it came to be known, was an original *BMW 718* bi-fuel rocket engine. Unlike the *HWK 509*, the pump required to pressurize the *C-Stoff* and *T-Stoff* rocket fuels was driven by the *BMW 003R* via a drive shaft from its accessory gear box. *Zborowski* had been working on a VTOL flying machine, too, when the war ended in May 1945.

How *Helmut von Zborowski* managed to find work in post war France is unknown because he belonged to that group of high-ranking *SS* members who were routinely prosecuted as war crimi-

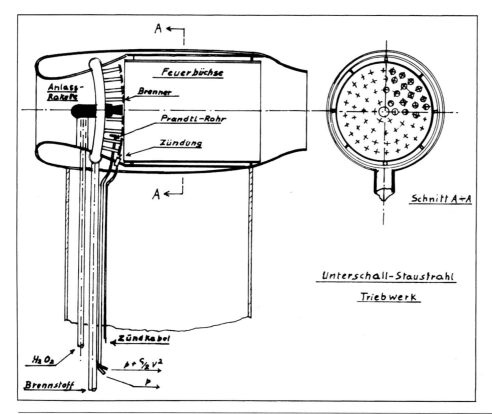

The ramjet engine requires air to enter at about 250 to 300 miles/hour for self sustaining operation. To bring the ramjet up to this operating speed a bi-fuel liquid rocket engine is required. In this pen and ink illustration of an *Otto Pabst* compact ramjet engine mounted on the rotor blade tip, its starting rocket engine is located in the air intake and is labeled *"Aniass-Rakete"*; the rocket unit is shown in black. Fuel to the rocket unit identified as H_2O_2 (hydrogen peroxide) is piped through the rotor blade from a storage tank in the *Triebflügel's* fuselage. A separate line for the catalyst (such as alcohol) would have run parallel to the H_2O_2 line.

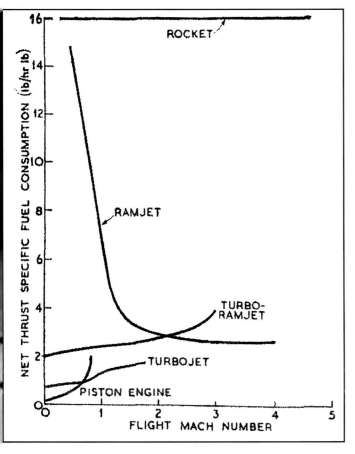

Vorgesehener Anbau des Triebwerkskörpers an FW 190 A-10

Ramjet engines have notoriously high fuel consumptions compared to piston or turbojet engines. Fuel consumption would have been about the same rate as a turbojet engine in full afterburner. In this pen and ink illustration is a graphic display of the net thrust of specific fuel consumption, and it can be seen how much fuel a ramjet requires at speeds up to Mach 1.0. It would make *Professor von Holst's Triebflügel* a very short range fighter/interceptor, and not a free roaming fighter.

Otto Pabst had plans to mount one of his miniature ramjet engines on the wing tips of a *Focke-Wulf Fw 190 A-10*. This pen and ink illustration shows how the *Pabst* ramjets would have been positioned on the *Fw 190 A-10* from an overhead view and a straight on view. It is not entirely clear if *Otto Pabst* ever modified an *Fw 190 A-10* with two of his compact ramjet engines.

This photo shows one of *Dr.-Ing. Otto Pabst's* miniature (compact) *Lórin*-type ramjet engines being wind tunnel tested in early 1944.

nals, *Horst Schnieder* told this author. Nor did he serve in the French Foreign Legion in Indo- China, as many other former *SS* did to escape prosecution. *von Zborowski* also had problems with *Ernst Heinkel* post-war. *Heinkel* and several others of his former *Heinkel AG* employees held patents on their proposed VTOL flying machine known as the *"Lerche"* from 1944/1945. No one in the *Heinkel* group would give *Zborowski* the needed patents when he sought to build his *"Coléoptère"* in France in the mid 1950s. When *Zborowski* did build his *"Coléoptère"* for the French aviation group *SNECMA* (*C.450.01*) it did not have mechanical rods or cables going to its controlling surfaces. Instead it had "wires" only which connected with small electric motors. This VTOL could claim the distinction of being the first aircraft to "fly by wire" of which we are hearing and seeing new technology designed into contemporary commercial and military aircraft.

A *Focke-Wulf Fw 190 A-10* of the type which *Otto Pabst* had intended to mount two of his experimental compact ramjet engines for flight testing.

Test Flight Experiences:
Going Vertical With The *XFV-1* and *XFY-1*

Test flight experiences from the only two American tail sitters is funny and humorous if it were not so dangerous. *Stephen Wilkinson* documented it in *Air & Space Magazine* article from October/November 1996. It stands to reason that the difficulties American test pilots had with the *Lockheed XFV-1* and *Convair XFY-1* would have also applied to the *Focke-Wulf Triebflügel. Wilkinson's* observations include:

• It was the riskiest military airplane *Lockheed* ever built...by their own admission;
• Takeoff was easy (*Convair*) but landing was something else. When you came in and pulled it up into the vertical, you were faced with three different configuration changes at the same time, all the time trying to figure out when to add power since you didn't want to

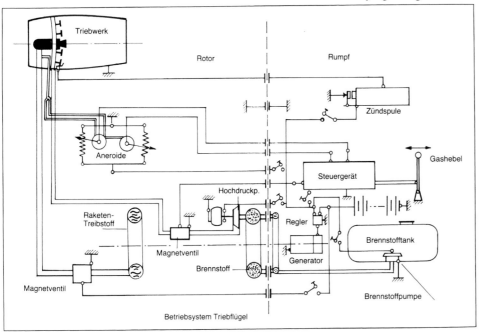

A pen and ink drawing featuring the entire operating system for one of *Otto Pabst's* ramjet engines for the *Triebflügel*.

***Otto Pabst*, in addition to designing a miniature ramjet engine for the *"Triebflügel,"* was also anticipating twin ramjet units for the *Hans Multhopp* designed fighter known as the *Focke-Wulf Fw Ta 283*. The *Ta 283* is shown from its port side. Scratch built fine scale model and photograph by *Günter Sengfelder*.**

Günter Sengfelder's **large, fine scale model of** *Hans Multhopp's* **proposed** *Focke-Wulf Fw Ta 283* **powered by twin** *Otto Pabst* **ramjet engines as seen from its nose starboard side.**

zoom up too high, because you had a hell of a time backing down again;

• When you pulled up, you had to rotate your seat (*Convair*). So you have a seat rotation change, a trim change, and a power change, and at the same your attention goes from your instrument panel to looking down back over your shoulder, because you can't fly the thing and follow the instruments. I still have a stiff neck from that;

• The terrible danger, however, was that at a descent rate of greater than 10 feet per second, the airplane (*Convair*) would suddenly tumble, totally out of control, most likely at an altitude too low for successful ejection.

• And the farther off the ground you were (*Convair*) , the more difficult it became to judge vertical" speed and altitude. Standing still at 500 to 600 feet, you would lose your depth perception;

• We couldn't look back over our shoulder (*Lockheed*) when flying the thing and judge height. We could practice landing on clouds all day, but this is the only airplane we ever built which we were afraid to fly ourselves in the final tests;

• As you would slow down and pull up-about the time you hit 30 degrees and 60 miles/hour, she'd want to start to roll and be quite unmanageable (*Lockheed*). At that point the prop wash was being deflected due to the extreme angle of attack and you weren't flying on the wing anymore, and the airplane was really stalled.

• We finally concluded that it was just foolish to risk the pilot's life (*Lockheed*). I had to write the letter to the Navy saying "Dear Navy: We'd like to quit this program. We should have said ("Hey, Navy, this whole idea is too stupid to continue.");

An overhead view of the *Otto Pabst* **ramjet powered fighter design by** *Hans Multhopp* **of** *Focke-Wulf Flugzeugbau.* **Scale model and photograph by** *Reinhard Roeser.*

A *Focke-Wulf* **proposed flying machine powered by twin** *Otto Pabst* **ramjet engines which are mounted on its swept forward horizontal stabilizer. Seen from its nose port side. Scale model and photograph by** *Reinhard Roeser.*

• Apparently, the "Tail Sitters" looked good on paper. What was never said (*Convair*) that the machine was a machine that nobody knew how to fly, that no simulator had proved would fly, and that no computer could promise would be controllable.

The Rotary Rocket

In 1999, the Rotary Rocket Company of Redwood Shores, California, test flew a flying machine they were calling the "*Roton*." This vehicle would take off vertically, powered by a spinning engine which contained 72 small rocket engines arranged around its base. After delivering a payload to low Earth orbit, the "*Roton*" would turn around, unfold its helicopter blades, and then reenter Earth's atmosphere, base first. Small hydrogen peroxide (H_2O_2) rocket engines located at the rotor tips, similar to the *Triebflügeljäger,* would keep the rotor blades spinning during the final flare to touchdown. The "*Roton*" was only flown twice and in the magazine *Air & Space* of February/March 2002, its only test pilot, former US Navy test pilot *Marti Sarigul-Klijn* with over 5,000 hours in 70 different type

Aircraft designer *Ludwig Mittelhüber* **of** *Focke-Wulf* **designed the** *Triebflügel's* **fuselage, while aerodynanicist** *Hans Multhopp* **designed its three rotors, as well as suggested their position on the fuselage.**

Otto Pabst **powered** *Focke-Wulf* **proposed flying machine as seen from beneath and its starboard side. Scale model and photograph by** *Reinhard Roeser.*

An overhead view of the three wide rotors on *Professor Erich von Holst's Triebflügel* spaced equal distance apart. Two of the four outrigger wheels can be seen.

of aircraft, described what it was like to vertically land the rotary rocket. These are his words.

The "*Roton*" blade tip H_2O_2 powered rocket engines required 4,300 pounds of 85 percent pure hydrogen peroxide. In 1999, this fuel cost $5.00 per pound and the "*Roton*" consumed about 1,000 pounds per minute of operation. This meant that there was only enough H_2O_2 for flights of 2 to 3 minutes duration.

Unlike a conventional helicopter, the "*Roton*" or rotary rocket did not have a tail rotor. What they learned was that it was very difficult to control rotor speed, because the rotor tip H_2O_2 rockets took more than one second to react to any change in the throttle. Also the rotor was unstable at operating speed. An increase in rotor speed (revolutions per minute or rpm) would cause the H_2O_2 rockets to produce more thrust, and that would make rpm go even higher. As rpm wound down, the reverse behavior occurred. *Sarigul-Klijn* found that he had to constantly work with the throttle to hold rotor speed constant. He found that it was very easy to overspeed the rotor. *Sarigul-Klijn* said that test pilots rate aircraft using what is called the *Cooper-Harper* scale, in which a "1" means an aircraft that's nice to fly and a "10" means an aircraft so difficult to fly that control will be lost during some portion of the flight. Eventually over 75 civilian and military test pilots flew the "*Roton*" simulator, most with at least 1,000 hours of flight time, and even after several hours of practice on the simulator, they all rated the rotary wing a "10." *Sarigul-Klijn* said that even after he had 10 hours of simulator practice, he could just barely complete a simulated vertical take-off and landing without crashing. He and others called the rotatory rocket's cockpit the "bat cave" because the view out of it was so limited. He said that they could not even see the rotor blades.

The rotary rocket offer many other challenges said *Sarigul-Klijn*. After each test run, the H_2O_2 rocket engines had to be removed from each of the four rotor blade tips and cleaned which also included the internal rotor blade plumbing. Otherwise they

Inboard of the three *Lórin*-type *Otto Pabst* miniature ramjet engines can be seen each of the starting rocket motors. This is one possible location and probably the best, because all the fuel lines would have to be thoroughly cleaned after each flight to get rid of any residue before the next flight. A time consuming job, perhaps taking several hours. Computer generated digital image by *Gareth Hector*.

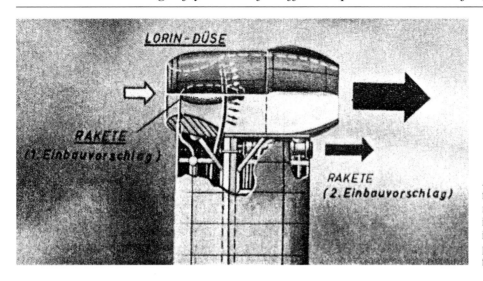

An illustration from *Focke-Wulf Flugzeugbau* featuring a second position for a starting rocket...right in the middle of the ramjet's air intake. Also featured is a rotor start motor buried in the trailing edge of the rotor blade. It appears the starting rocket locations were not yet fixed by war's end.

A close up view of the *Otto Pabst Lórin*-type ramjet engine with its rotor starting rocket inboard the ramjet engine. Computer generated digital image by *Gareth Hector*.

might unexpected stop due to catalyst debris left during shutdown. Cleaning these four H_2O_2 rocket engines took most of the day. *Sarigul-Klijn* said of the rotary rocket, like most helicopters, it was inherently unstable in hover and because it was tall, its instability was exaggerated. For example, when the rotary rocket wobbled side to side, it also wanted to turn left or right. A hard sideways landing could cause the rotatary rocked to fall over.

The "*Roton*" was unable to attract investors and the Rotary Rocket Company abandoned the concept as being too technically ambitious. It now stands abandoned in its Mojave desert hangar.

Reflections On The Driving Wing "*Triebflügeljäger*"
The following is from a speech titled "Driving Wing (*Triebflügeljäger*)" from about 1954. Author unknown, however, it was probably *Dr.-Ing. Theodor Zobel*, who came to the United States via "Operation Paperclip." It appears that the speaker was appearing before an aviation engineering design group and the speaker appears to know a great deal about this former *Focke-Wulf Flugzeugbau* project.

Dr.-Ing. Theodor W. Zobel, Otto Pabst's research colleague on ramjet engines at *LFA-* Braunschweig and the former chief of its high-speed aerodynamic section of the Aerodynamic Research Center, came to the United States postwar in Operation Paperclip. He developed the Schlieren-Interferometer, an optical measuring device to measure the flow of air around an airfoil without disturbing the flow. This was one of the most important contributions to supersonic research. His work, post war at the Wright Air Development Center, Dayton, Ohio is believed to have saved the United States several years of expensive research time. *Theodor Zobel* died in Cincinnati, Ohio. The speech of *Dr.-Ing. Zobel,* its likely author, is reproduced here word for word.

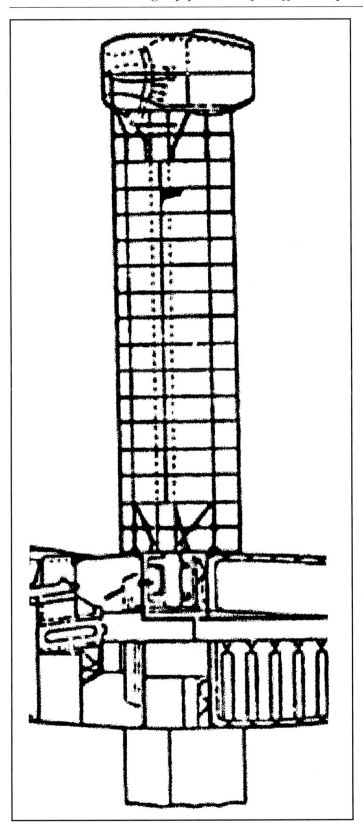

A drawing featuring some of the internal structure of the rotor blade from ramjet engine to the *Triebflügel's* fuselage, and the plumbing to and from the *Otto Pabst* miniature ramjet engine.

Introduction

"In postwar reports on aircraft development during the last war you may have noticed some pictures of vehicles which must have appeared very strange - at least to someone who held well established opinions on how an aircraft ought to look. One of these freaks was a *Focke-Wulf Flugzeugbau* project authorized by its director *Kurt Tank* called *Triebflügeljägerjäger* or driving wing fighter. *Trieb* in German means a drive, an urge, an inner compulsion such as love or hunger, and *flügel* means wing. *Jäger* means fighter. In the colorful language of modern officialdom, we would call it a VTOL aircraft. I don't expect that any one of these terms is particularly revealing. Therefore, I would like to show you a picture so that you have an idea of what I am going to talk about.

History

It all began with the attempts of a zoologist in Göttingen, *Dr. Erich. von Holst*, to understand more about the mechanics of animal flight. Besides the inborn curiosity of the real scientist, this man had the gift of building rather intricate model planes and making them fly. Thus he constructed quite a number of rubber band-powered artificial birds whose flight looked impressively natural.

When *von Holst* compared his birds or most of the flying animals with the man-made airplanes of that period, he noticed an important difference: those birds had only one set of organs for propulsion and lift generation in flight, whereas, in most aircraft these two basic functions of flying are handled by completely different devices. The animals were even more versatile: their ability to create enough lift was not limited by a minimum flight speed.

Together with *Dr. Dietrich Küchemann* from the *AVA*-Göttingen he began to analysis the principles of animal flight with an eye on the possibility of applying some such principle to aircraft. It was fairly obvious that the flapping motion of bird wings was not a technically desirable solution to the flight problems, partly because of the very high forces developed in such slow up and down motions, partly because of the slow oscillations of the body of such and airplane. There was, however, one animal which seemed particularly free of these body oscillations: the well-known dragonfly. This large insect uses two sets of wings one behind the other which swing 180 degrees out of phase, for example, one set goes up if the other goes down. Thus, *von Holst* built a model of a dragonfly which showed the same effect. Since he found two complete sets of movable wings a bit too complex he then began to simplify his artificial dragonfly. First, he changed to two rigid wings, one behind the other, which were oscillating about hinges parallel to the body axis, again 180 degrees out of phase. The next step was fairly obvious: the oscillation of the two wings was replaced by the technically much simpler complete rotation about the same fuselage axis. After some experiments with the size and location of tail-surfaces *von Holst* came up with a model plane which flew in any direction in space, vertical, or horizontal, or at any odd flight path angle with a contra-rotating airscrew acting as propeller and lifting device.

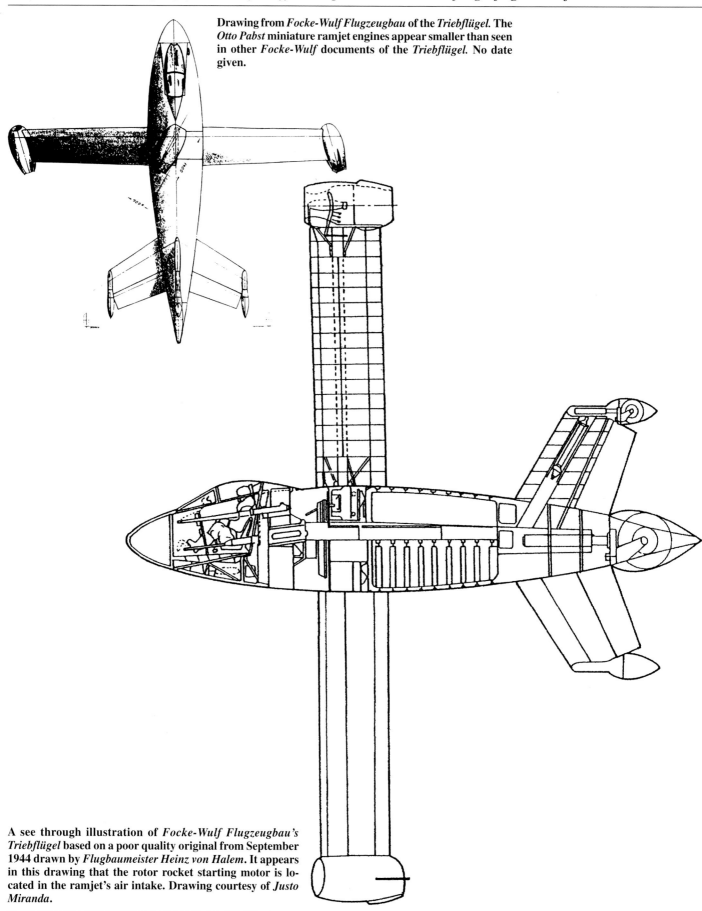

Drawing from *Focke-Wulf Flugzeugbau* of the *Triebflügel*. The *Otto Pabst* miniature ramjet engines appear smaller than seen in other *Focke-Wulf* documents of the *Triebflügel*. No date given.

A see through illustration of *Focke-Wulf Flugzeugbau's* *Triebflügel* based on a poor quality original from September 1944 drawn by *Flugbaumeister Heinz von Halem*. It appears in this drawing that the rotor rocket starting motor is located in the ramjet's air intake. Drawing courtesy of *Justo Miranda*.

This was the flying machine which *Dietrich Küchemann* analyzed particularly with regard to its performance potential. He found it did not compare badly with orthodox airplanes of the same period. This analysis was mostly based on the limiting case of a slowly rotating wing for which the results of *Professor Ludwig Prandtl's* airfoil theory could be used. Yet, a few points seemed not too good; there was, for example, the need for a two-speed reduction gear between the large contra-rotating propeller-wing and the engine. Also, the power requirements were quite remarkable if decent speed capabilities were expected; for example some 7,500 horsepower

for an airplane weighing about 22,000 pounds. This was, of course, beyond the state-of-the-art at that time. However, it is interesting to note that the numbers represent roughly the power situation for the two VTOL aircraft presently developed by *Lockheed* and *Convair*. Another step seemed still necessary in order to arrive at some practical airplane configuration. This came from a different direction: at the *Focke-Wulf Flugzeugbau*, we had been interested in the development problems of turbojet engines, for our own use. However, the *RLM* or German Air Ministry under pressure from the engine industry proper forbade our activities in this field with the exception of ramjets, for which good applications were not so easily found at that time. Here seemed to be one. If we put our ramjets (which were quite successfully developed by *Dr.-Ing. Otto Pabst* based on the research of *Eugen Sänger*) on the tips of the

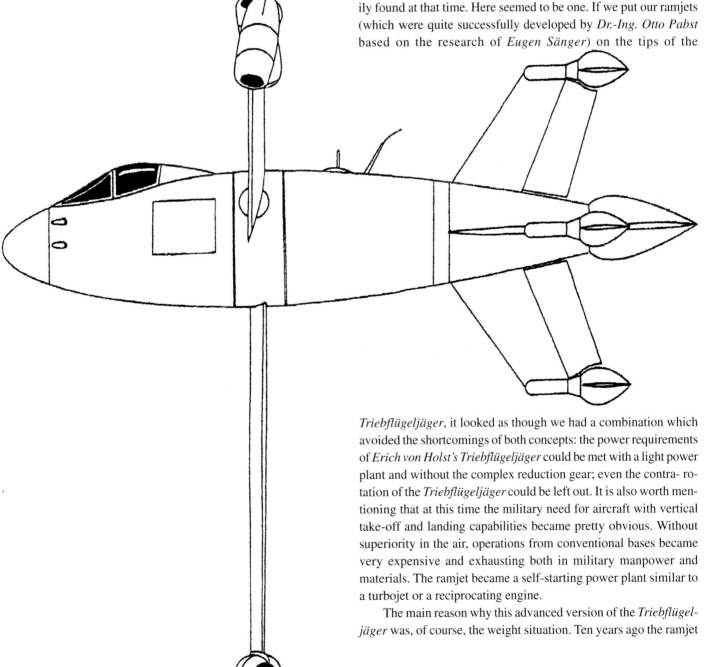

Triebflügeljäger, it looked as though we had a combination which avoided the shortcomings of both concepts: the power requirements of *Erich von Holst's Triebflügeljäger* could be met with a light power plant and without the complex reduction gear; even the contra- rotation of the *Triebflügeljäger* could be left out. It is also worth mentioning that at this time the military need for aircraft with vertical take-off and landing capabilities became pretty obvious. Without superiority in the air, operations from conventional bases became very expensive and exhausting both in military manpower and materials. The ramjet became a self-starting power plant similar to a turbojet or a reciprocating engine.

The main reason why this advanced version of the *Triebflügeljäger* was, of course, the weight situation. Ten years ago the ramjet

A fully skinned illustration of *Focke-Wulf Flugzeugbau's Triebflügel*. Courtesy of *Justo Miranda*.

was the only power plant besides the rocket with gave plenty of power at a moderate weight. Turbojets at that time were still weighing over 0.6 pound per pound of thrust, turboprops were not yet developed, and reciprocating engines came at considerably more than 1.0 pound per horsepower.

The *Triebflügeljäger* was intended as an interceptor aircraft of about 11,000 pounds weight. The propeller diameter was about 33 feet; the small diameter ramjets had a diameter of about 2 feet; they were tested in the open jet high-speed wing tunnel of *LFA*-Braunschweig, and delivered a positive thrust of the expected or-

der of magnitude up to Mach numbers of 0.9 [667 miles/hour]. The design studies which were mostly done by *Focke-Wulf's Flugbaumeister Heinz von Halem* covered the usual problems of a tail-sitter airplane: seat arrangements of the pilot, the landing gear, for example, ranges for stable flight, and so on. *Frauline Dr.-Ing. Irene Ginzel* from *AVA*-Göttingen, analyzed more thoroughly the forces on the rotating wing system in flight situations in between the two extremes studied by *Kuchemann*, and computed the optimum twist distribution along the blade and the pitch angles required. These calculations took the helical vortex pattern of the flow field behind the *Triebflügeljäger* into account.

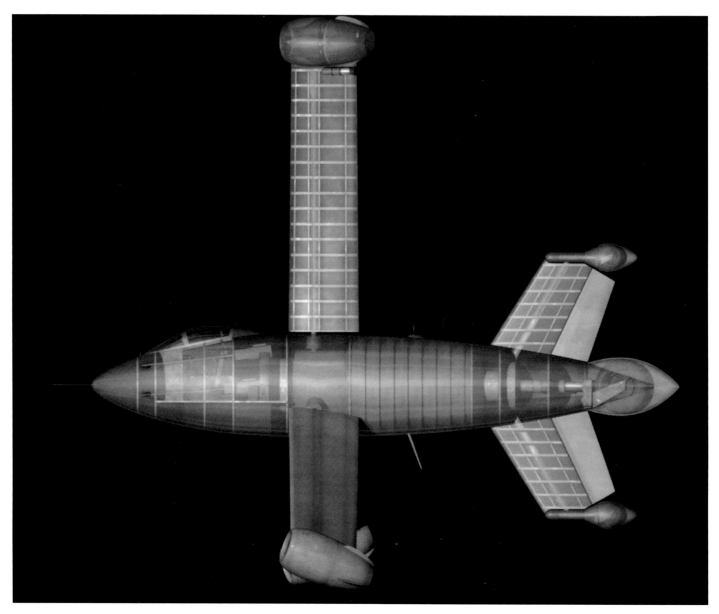

A see through computer generated digital image of the *Triebflügel* by *Gareth Hector*. Beginning in the nose were the four cannon...two mounted each side the cockpit. Immediately aft the cockpit was the rotor attachment ring. Aft the attachment ring was the fuel tank for the flying machine's ramjet engine. There would have been a small tank for the rocket engine's fuel—most likely hydrogen peroxide—as well as one for the oxidizer—alcohol. Aft the fuel tanks was the massive hydraulic cylinder for the fuselage mounted main wheel and its clamshell aerodynamic covers. Its four outrigger wheels were attached to the four aft stabilizers. Each outrigger wheel had its own clamshell cover. In all, the *Triebflügel* was a very aerodynamic design.

By the end of the war all these effects were still mainly in the preliminary stage. We were aware of the many problems which were practically unsolved.

Performance Outlook of the *Triebflügeljäger*

If we want to know what the performance possibilities of a flying machine such a *Triebflügeljäger* one would have to study mainly two cases: hovering flight (V = O) and high speed horizontal flight. At very low speeds the *Triebflûgeljäger* would act as some kind of thrust amplifier. The thrust of a propeller at sea level conditions is given by a relation T = const.(P,D) 2/3, P being the power absorbed by the propeller and D its diameter. With the ramjet, thrust T, acting essentially tangential on the blade tips the power is P = WRT; the circumferential velocity WR is usually limited by centrifugal forces in the rotor blade. I thought 660 feet/second [200 meters/second] was a possible value. If we consider WR as constant we have:

$$T = const.\left(T_j D\right)^{2/3} \text{ or}$$

$$T = const.\left(P_l D\right)^{2/3}$$

I = const. if we measure the thrust in pounds and D in feet the constant is about 25, so that

$$T, 3\sqrt{T^1/D^2}$$

T ∝ 25D For the *Focke-Wulf* aircraft we expected a thrust of about 13,000 pounds during the

$$T_1 \sqrt{T}$$

take-off phase which meant an amplification factor T in the order of 7. The SFC of the ramjet was about 4#[1] HR at the tip speed considered so that the SFC of the resulting thrust reached a value of about .6#[2] HR which is better than most jet engines although not as good as a turboprop.

With increasing flight speed we observe a change in this pitch angle at the blade tip which means that more and more of the jet thrust goes directly into propulsion and propulsive efficiency of the airscrew becomes less important. More critical is the lift-drag relationship of the *Triebflügeljäger*.

At first glance one would assume that the distribution of the lift over the whole propeller area results in a small induced drag. The theoretical induced drag of a lifting system that is circular in its front view was compound by *Professor Ludwig Prandtl-Göttingen* many years ago at just one half the induced drag of a

A cut away illustration from the *Focke-Wulf Flugzeugbau* featuring the internal structure of the rotors and rear stabilizers.

wing with the span equal to the diameter of the lifting system. This is, however, a much too optimistic case. The lift is not spread evenly over the whole propeller plane, but is concentrated on the blades. In the case of a slowly rolling wing (2 rotor blade propeller) this results in an induced drag twice as high as that of the equivalent fixed wing as already shown by *Dr.-Ing. Küchemann.* The reason is simple: in most positions of the blade only part of the tangential force on the blade contributes to the lift, but the instantaneous induced drag depends on the whole lift force on the blade; but the instantaneous induced drag depends on the whole lift force on the blade; and the instantaneous lift-values of the blade vary between 0 in the vertical position and twice the mean value in the horizontal position, which means that the peak values of the instantaneous induced drag of 4 or 5 times as high in the equivalent fixed wing. For a rotating lifting system with more than 2 (rotor) blades the relations should gradually improve. It can be shown that a four (rotor) blade system might ought to have the induced drag of the equivalent fixed wing.

Experiments with propellers which were, of course, not designed to develop lift forces efficiently show rather disappointing

results. The results are taken from tests by *J.F. Runkel*, NACA-ARR 88, June 1942. Since the induced drag could appear as an increase in power required, or as a loss in thrust it was evaluated from:

$$\frac{\Delta P - \Delta T}{V}$$

$$\frac{Di}{\varrho \dot{V}^2 D^2} = \frac{\Delta C_p}{J^3} - \frac{\Delta C -}{J^2}$$

which is plotted over the square of the lift (normal force) number.

It can be seen easily that most of the experimental values are twice as high as this seemingly pessimistic theory. The reason for all this discrepancy is not too simple: it may be that the unsteady flow effects due to the cyclic change in lift during each rotation increases the induced drag still more; more probably the lift distribution along the blades was different from the ideal elliptical, and the blade sections were too thin to develop some nose thrust.

Based on such tests the empirical effective span of such a rotating lifting system must be assumed to be about equal to the propeller radius. This would affect the high altitude and range possi-

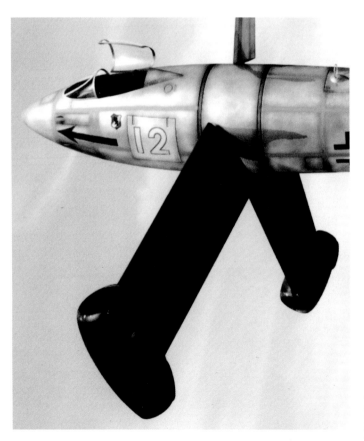

The *Treibflügel* as seen from its port side. Notice the massive ring surrounding the fuselage aft the cockpit for securing the three rotor blades and the *Otto Pabst* miniature ram jet engines. Notice, too, how the cockpit canopy opens to starboard. Computer generated digital image by *Gareth Hector.*

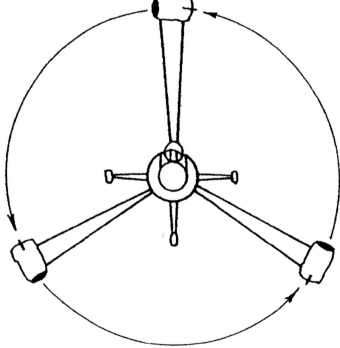

A drawing featuring the overall radius of the three rotor blades, including their *Otto Pabst* miniature ramjet engines. The rotation of the rotor blades was in a clockwise direction when viewing the *Triebflügel* head on. Drawing courtesy of *Justo Miranda.*

A port side view of the *Triebflügel* standing on the ground looking up. Notice the clear view of the rotor blade with its twist and its *Otto Pabst* miniature ramjet engine located out at the rotor's tip. Scratch built fine scale model and photographed by *Günter Sengfelder*.

bilities of *Triebflügeljäger* aircraft considerably. For the *Focke-Wulf* aircraft we would expect, therefore, with a drag area of 3 square feet with a wetted area of about 300 square feet, a lift/drag ratio of about 8.5 and, therefore, a range factor nautical miles compared to

$$\frac{V}{SFC} \times \frac{L}{D} = 1000$$

between 5,000 nautical miles and 10,000 nautical miles for most other jet aircraft. The thrust area of the 3 powerplants had been 2 square feet per unit so that with the fuel weights planned a

range of 250 nautical miles should have been expected. The ceiling of that aircraft at combat weight would be near 40,000 feet altitude. It is not unlikely that some development effort would improve these values considerably.

Are There Changes We Could Have Done to Improve The *Triebflügeljäger* Concept?

The development of the *Triebflügeljäger* until the end of the war had certainly not reached the limit of what may be possible with this kind of aircraft. There are at least two lines of further development which promise some progress:

A rare photo featuring a pair of twin rotor *Treibflügels*. The twin rotors were the early design ideas. Notice the pilot's cockpit is situated way aft, based on the cruciform tail, which is much different from the three rotor version that came to be the preferred *Triebflügel* arrangement. The two rotor blade *Triebflügel* seen on the left appears to have its rotors locked, thus behaving like a normal wing. It is not clear if the *Triebflügel* would have flown in horizontal flight with its rotor blades locked in place, or if the rotors continued to rotate around the fuselage. If this were the case the rotating blades would likely have created a tremendous amount of drag.

An older illustration featuring the three rotor blade *Triebflügel*. It appears that a pilot is ready to climb into the *Triebflügel* on the right, while the *Triebflügel* on the left appears to be crossing over the air station at full horizontal speed.

1. Better fuel economy at subsonic speeds through the replacement of ramjets by turbojets. This could improve range and endurance by a factor of about 4. The turbojets could be located at the blade roots with only the exhaust piped into the tips if mechanical problems due to centrifugal and Coriolis forces were too severe;

2. Try for supersonic speed range. Here ramjet become more and more efficient. To make it possible the blades have to be very thin since sweeping seems hardly possible. It is not unlikely that the body shaping in the NACA area rule fashion will bring some improvement.

Comparison With Other VTOL Possibilities

In order to say whether it is worthwhile to do some more development work with airplanes of this class one has to compare the *Triebflügeljäger* with other aircraft which have vertical take off and landing possibilities.

There is first the conventional helicopter. The main difference is the always almost vertical fan axis. This limits the speed severely; on the other hand it eliminated the need to go through a large range of angles of the body axis. For the transportation of many goods this is an absolute necessity, for example, at a tail sitter for the transport of passengers sounds hardly reasonable. Range capabilities are of the same order of magnitude for both types; altitude performance is better with the *Triebflügeljäger*.

Digital image by *Mario Merino*.

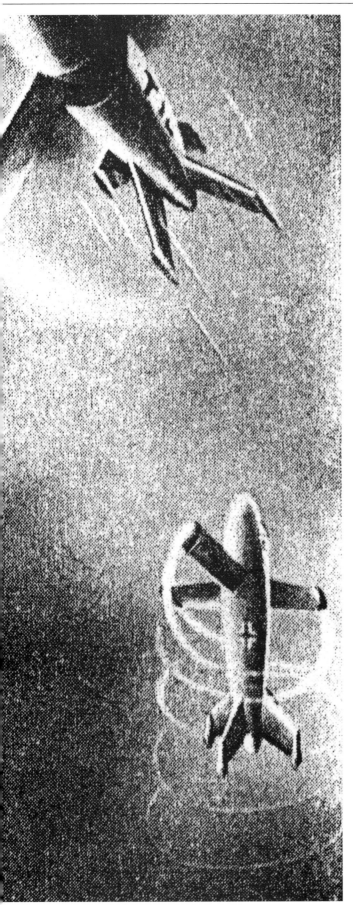

A drawing of a *Focke-Wulf Triebflügel* by *Carlo Demand* seen lifting off from a clearing in the forest. Courtesy: *Flug Revue & Flugwelt*, December 1981.

Another type of VTOL aircraft which, like the *Triebflügeljäger* combines lift and propulsion in one organ is the French flying barrel (*Zborowski's SNECMA C.450.01 Coléoptère*). This has been suggested fairly often; it consists of an overgrown jet or ducted fan power plant whose large cowling acts as a wing. Performance with this kind of an airplane should have the same limitations as the *Triebflügeljäger* only more so because its diameter can't grow even into the order of magnitude of the propeller radius of the *Triebflügeljäger*.

If we drop the idea that propulsion and lift have to come from the same devices we get aircraft with power plants big enough to provide the lift for the hovering flight. These power plants can use a propeller like the *XFY-1* and *XFV-1* or they are of the jet family. The propeller planes are roughly in the same performance class as the *Triebflügeljäger*. The combined weight of wing plus power plant plus propeller should exceed the weight of the rotating part of the

A poor quality older illustration of a three rotor *Triebflügel* appearing to be making a vertical lift off with the exhaust from the *Otto Pabst* ramjet engines making spirals as it gains alitutde.

Courtesy: Huma Modells.

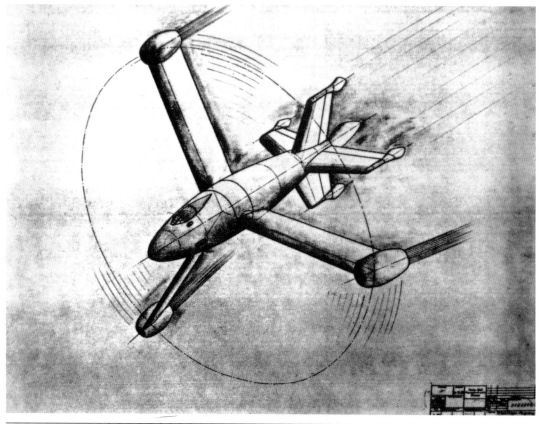

An illustration by *Focke-Wulf Flugzeugbau*. What is very interesting about this drawing is that the *Triebflügeljäger* is shown with its three rotor blades locked for horizontal flight. It is not entirely clear if during horizontal flight the *Triebflügel's* rotors were locked or continued to rotate around the fuselage, as they did during vertical lift off and landing.

| | Geheime Kommandosache | | -2- | Gehcime Kommandosache |

Triebflügeljäger mit Lorinantrieb.

Allgemeine Daten

Verwendungszweck:	Einsitziger Jäger mit Druckkabine.
Bauart:	Triebflügelflugzeug mit Lorinantrieb.
Festigkeit:	$n_1 = 6,0$ bei $G = 5,175$ t
Triebwerk:	3 Lorindüsen in Fw-Bauweise
Baugrößen:	Triebflügel-Blattfläche $F = 16,5$ m²
	-Schraubenfläche $F = 80$ m²
	Spannweitenhälfte $b/2 = 5,0$ m
	Streckung $\lambda = 9,1$
	Seitenleitwerksfläche $Fs = 5,0$ m²
	Höhenleitwerksfläche $Fh = 5,0$ m²
	Größte Länge $L = 9,15$m
	Größte Breite $B = 11,50$m
	Flächenbelastung max. $G/Fmax = 314$ kg/m²
	Flächenbelastung min. $G/Fmin = 212$ kg/m²
Besatzung:	1 Mann
Fluggewicht:	Größtes Startgewicht $Gmax = 5,175$ t
	Kleinstes Landegewicht $Gmin = 3,500$ t
Fahrwerk:	1 Landerad 780 x 260 mm
	+4 Rollräder 380 x 150 mm
Landehilfe:	Für Triebwerksausfall 3 Walter-Geräte von je 300 kg Schub in den Lorindüsen.
Kraftstoffanlage:	Geschützte Rumpfbehälter von 1500 kg Inhalt
Bewaffnung:	2 MK 103 mit je 100 Schuß
	2 MG 151 mit je 250 Schuß
Panzerung:	Übliche Jägerpanzerung (100 Winkel von vorn)

Leistungen:

Höchstgeschwindigkeit	0 km Höhe	1000 km/h
	7 km "	900 km/h
	11 km "	840 km/h
	14 km "	840 km/h

Steiggeschwindigkeit	0 km Höhe	125 m/s
	7 km "	50 m/s
	11 km "	20 m/s
	14 km "	7 m/s
	15 km "	2 m/s
	15,5 km "	0 m/s

Steigzeiten auf	1 km Höhe	8,2 sec
	2 km "	16,8 sec
	4 km "	39,5 sec
	8 km "	1,8 min
	12 km "	4,5 min
	14 km "	7,4 min
	15 km "	11,5 min

Reichweite	0 km Höhe	650 km	bei $v_R =$ 925 km/h
	4 km "	930 km	870 km/h
	8 km "	1350 km	600 km/h
	12 km "	2000 km	725 km/h
	14 km "	2400 km	675 km/h

Flugdauer	0 km Höhe	0,7 h	bei $v_R =$ 925 km/h
	4 km "	1,0 h	870 km/h
	8 km "	1,5 h	800 km/h
	12 km "	2,6 h	725 km/h
	14 km "	3,4 h	675 km/h

Brennstoffverbrauch im Steigen auf	4 km Höhe	80 kg
	8 km "	170 kg
	12 km "	260 kg
	14 km "	340 kg

Gewichte:

Rumpfwerk	475 kg	
Fahrwerk } kombiniert	250 kg	
Leitwerk }	225 kg	
Steuerwerk	60 kg	
Tragwerk	575 kg	
Lorindüsen Triebwerksanlage	240 kg	
Walter-Geräte	125 kg	
Schützer (geschützt)	350 kg	
Ausrüstung (fest)	225 kg	
Panzerung	175 kg	
Waffen 2 MK 103 + 2 MG 151	500 kg	
Rüstgewicht	3200 kg	3200 kg
Betriebsstoff Lorindüsen	1500 kg	
Walter-Geräte	90 kg	
Munition 200 Schuß MK 103	170 kg	
500 Schuß MG 151	115 kg	
Besatzung	100 kg	
Zuladung	1975 kg	1975 kg
Fluggewicht:		5175 kg

Bad Eilsen, den 15.9.44

Photocopy of a specifications sheet from *Focke-Wulf Flugzeugbau* titled *"Triebflügeljäger mit Lorinantrieb"* and dated September 15th 1944.

Triebflügeljäger. On the other hand the power plant efficiency can be better.

In the long run the turbojet powered VTOL aircraft has probably the best chance. Specific weight of turbojet engines has come down considerably, and we can look forward to engines which weigh less than 1/10 of their sea level static thrust. This is why I feel that it is hardly worth while to go to deeply into the development of some of the other more unconventional looking aircraft configurations such as the *Triebflügeljäger* or the French flying barrel of *Zborowski* although they are technically very interesting.

I did not mention much of the stability and control problems of the *Triebflügeljäger* and other VTOL aircraft. They have been studied intensively but the final proof namely the actual flight is still missing. The normal force of the propeller in non-axial flow is used as lift in the *Triebflügeljäger* is, of course, still there when we don't need it at all, rarely, in the hovering flight condition. There it is a very destabilizing force which must be compensated by artificial stabilization. The same is true for any VTOL aircraft using a propeller; here again the jet is considerably simpler. It also shares with the propeller driven VTOL aircraft the problems resulting from wind effects on the slipstream in the tail area.

The vertical take off and landing is indeed only the extreme case of a short take off and landing. Therefore, the VTOL aircraft proper must also be compared with aircraft of a more conventional type which have efficient devices to fly slowly or to cut down the acceleration and deceleration phase in take off and landing. Such devices as boundary layer control, catapults, and zero length launchers, light weigh boost power and thrust reverses on the regular power plant or drag chutes and elastic landing barriers are going to improve the competitive position of the conventional aircraft quite a bit. It is also worth noting that the early pioneer phase of turbojet power development is over. Even without any of the above mentioned devices there ought to be a considerable improvement in the take off features of regular aircraft due to the development of spe-

53

cifically lighter engines. We will soon regard most of these early turbojet aircraft as what they are: extremely underpowered and ill compromised with over-emphasis on top speed. In the field of military aircraft there is an ever growing interest in more altitude and speed. The engines and the airplane configurations which give us this increase will produce considerably shorter take off and landing distances almost automatically. Some moderate additional effort in the form of boundary layer control and lift from thrust producing devices can bring almost VTOL capabilities for many types of fixed wing aircraft in the future."

It was mid August 1944 before *Otto Pabst's* first operational ramjet was ready for testing. Prior to its testing with propane gas, it was taken to the wind tunnel of *LFA* at Braunschweig. It could not be accommodated despite pressure put on them by *Oberst Siegfried Knemeyer*, the boss of the *RLM's* technical office. Even *Pabst's* ramjet engine had to wait for its first operational tests because the facility which manufactured propane gas, the Leuna-Werks, had been recently bombed by Allied bombers. Propane gas for *Otto Pabst's* ramjet engine would not available until the facility had been restored. Even then it was not known for certain when the Leuna-Werks would be restored sufficiently to begin the production of propane gas.

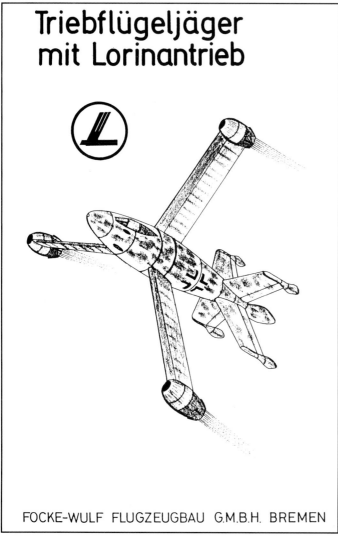

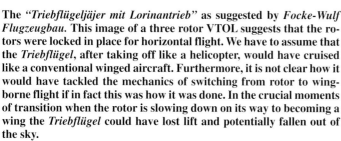

The *Triebflügel* seen from its port side shortly after lift off. This point interceptor was to stand vertically on the ground supported by four tail fins, each of which had a smaller outrigger wheel at its tip. The main landing load was to be taken by a single main wheel at the base of the fuselage. During flight, as seen in this digital image by Gareth Hector, all wheels were to be enclosed by streamlined, tulip-shaped pods.

The *"Triebflügeljäger mit Lorinantrieb"* as suggested by *Focke-Wulf Flugzeugbau*. This image of a three rotor VTOL suggests that the rotors were locked in place for horizontal flight. We have to assume that the *Triebflügel*, after taking off like a helicopter, would have cruised like a conventional winged aircraft. Furthermore, it is not clear how it would have tackled the mechanics of switching from rotor to wing-borne flight if in fact this was how it was done. In the crucial moments of transition when the rotor is slowing down on its way to becoming a wing the *Triebflügel* could have lost lift and potentially fallen out of the sky.

It was mid September 1944 that *Dr. Pabst* decided to switch fuels to hydrogen gas from propane. Even finding enough hydrogen gas in bottles took time even though the *Luftwaffe* at Orainenburg was searching throughout Germany for hydrogen. It was not until the middle of January 1945 that *Pabst* and his assistants were able to test the burners of his experimental ramjet engine in a cold chamber device at Oranienburg, producing a thrust of 0.705 kilograms/second. The next development step sought by *Dr. Pabst* was a flight trial for the purpose of clarification of his ramjet's general flight behavior. He anticipated to attach ramjets to the wing tips of a *Focke-Wulf Fw 190* fighter aircraft. Drawings were made on how to carry out the necessary modifications however, it was not carried out due to a shortage of *Focke-Wulf Fw 190s* as well as the ability to carry out such tests at altitude due to the constant Allied fighter presence.

Otto Pabst's development trial program, which had been set up in December after an order by *OKL* to construct four of the experimental ramjet for field testing by August 1945 came to an abrupt

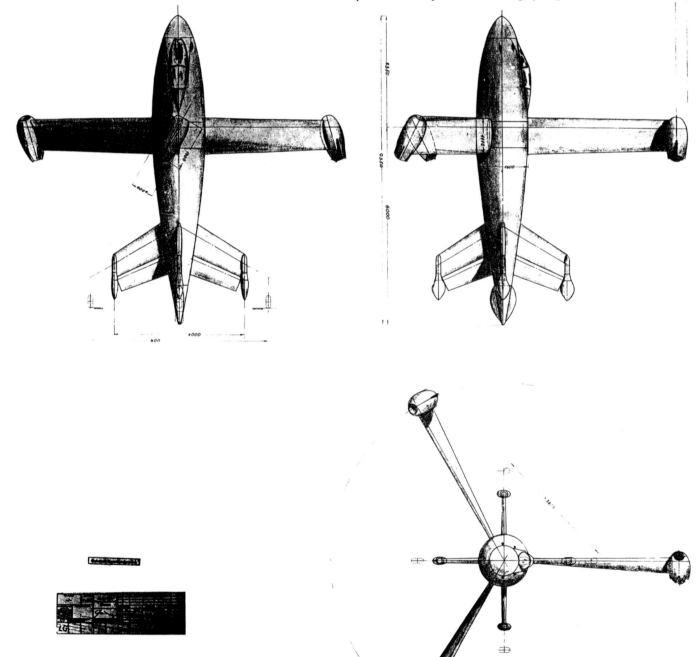

A drawing of the *Triebflügeljäger* from *Focke-Wulf Flugzeugbau* from mid-1944. The idea for a VTOL flying machine was becoming increasingly popular due to *Nazi* Germany's growing inability to launch conventional fighters, such as the *Messerschmitt Me 262A-1a* and *Focke-Wulf's* own *Fw 190*.

end with the Allied occupation of Braunschweig and Bad Eilson.

The research ramjet research work of *Dr. Zobel* and *Dr. Pabst* was evaluated by both the American and English research institutes. It was not declassified until January 1955.

In the early summer 1944, however in *Focke-Wulf*, a hunter design, that many specialists designed the "*Egg of Columbus*" began to ripen. It was appearing that the *Focke-Wulf* people were beginning to believe that the *Lórin* powered *Triebflügeljäger* was a do able project. *Hans Multhopp*, the authoritative designer of the *Focke-Wulf Ta 183* and *Otto Pabst* had been working together for the rotor placement and pilot's cockpit placement on the fuselage and the starter engine question whose pipes and controls had to go on the inside of each rotor to the ramjet engine. However, it was the final placement of the rotors....first on the nose, then aft, and finally

about one-third of the fuselage length back from the nose that the designers believed that they had indeed designed the "*Egg of Columbus*."

The rotor was moved by the thrust from the ramjet engine on the rotor tip. Ignition was immediate upon reaching a specific forward speed and increased with increasing dynamic pressure until after about one minute developed sufficient rotation to provide the lift needed for vertical take off.

By May 1944 the detailed drawings had been completed. At the same time the placement of the rotors involved three locations on the fuselage. Its location had to be in an area where the fuselage had the widest diameter due to the requirement for reduction gears and accessories. Placement of the machine cannon was also a requirement. Placement of the pilot's cockpit went through several design studies. It was believed that a cockpit place behind the ro-

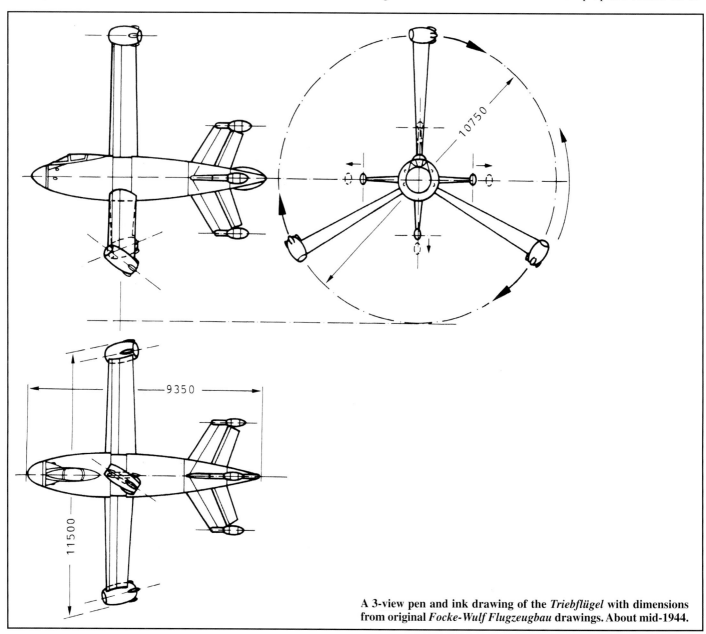

A 3-view pen and ink drawing of the *Triebflügel* with dimensions from original *Focke-Wulf Flugzeugbau* drawings. About mid-1944.

Computer generated digital image by *Marek Rys*.

tors was disadvantageous. A location in front of the rotors was deemed to be best as was the location of the machine cannon along the cockpit fuselage sides.

A JABO-variant was considered the natural way to start the ramjet engines. The start method with rockets was to be investigated such as maintenance, expenses...rockets with different burn times, however, perfection of a ramjet engine was still considered a long way off so that JABO starting rockets was to wait until the ramjet engine was perfected.

In the arrangement/placement of the rotors (c) could be done by the usual means of attaching a propeller to an engine. In (a) and (b) required a great deal more support around the entire fuselage and complicated engineering.

Professor von Holst's model flew with stability. He later reported it at the time in the 1942 annual of Aeronautical Research in great detail over its success. *von Holst* would later demonstrate his paper and wood model before *Kurt Tank* and his *Focke-Wulf Flugzeugbau* designers at Bad Eilson so that *von Holst* began to busy himself with the aircraft designers there.

The formula shows the considerable influence of the size, length to diameter. For the *Lórin* engine, the task at *Focke-Wulf Flugzeug*bau was holding the length as short as possible. In addition, *Fangdiffusor* as described by *Dr. Küchemann* of *AVA*-Göttingen with the internal diffusor. *Küchemann* also helped design the nozzles for cooler operation. After the laying down of the principles of the

A pair of *Triebflügels* seen making their transition to horizontal flight. The artist presents these two *Triebflügels* moving along with their rotor blades rotating about the fuselage. We do not know how the transition would have occurred.

The two *Triebflügels* are presented in horizontal flight. How far the ramjet powered *Triebflügel* could have flown horizontally is unknown. What is known is that the ramjet engine consumes fuel at a rate similar to a turbojet engine with full afterburner operating. Other wise, this VTOL would have had a very specific role...vertical take off to altitude to meet and do combat with the bombers and then return immediately back to its launch site.

short length ramjet engine by *Otto Pabst*, *Dr. Zobel* had nothing to do with the overall design. A model of the ramjet was taken to the wind tunnel A 4 at *LFA*-Braunschweig which was under the direction of *Adolf Busemann* who was also subordinate to this whole process. Other investigators included *E. Wolfhard Schmidt* and *Professor Albert Betz* in the *AVA*-Göttingen.

Propulsive Duct (*Luftstrah Antrieb* or *L-Triebwerk*)

A research program to test the principle of the propulsive duct (*Lórin* duct) had started in Germany as early 1936. One of the first was the *Helmuth Walter Werk* at Kiel (*HWK*). Research on the propulsive

Flight control of the *Triebflügeljäger* was to have been accomplished by means of control surfaces at the trailing edges of the four tail fins. For flying in an horizontal position the tail might have been depressed slightly to direct part of the thrust force into a lift force. Digital image by *Gareth Hector*.

No development work is known to have been carried out on *Focke-Wulf Flugzeugbau's Triebflügeljäger* concept, and the viability of the design is still a matter of speculation. Digital image by *Marek Rys*.

Families of ALTERED WARS

Luftwaffe: 1946

No.1
SPECIAL
FEB. 1997
$2.95 U.S.
$4.10 Can.

V2

TRIEBFLÜGEL Special

Ted Nomura/Ben Dunn

duct involved the combined efforts of a great many individuals and the scheme was beginning to look attractive as a practical method of aircraft propulsion starting in late 1944 and early 1945. The first documented American *Lórin* duct (they actually called their propulsive duct a "heat model") was investigated by *John Becker* and teammates of the NACA Langley, Virginia in their 8 foot high speed tunnel in February and March 1941.

Some of the original propulsive duct designs dating back to 1936 had a combustion chamber of very large diameter so that the gas velocity at the combustion chamber entry would be low and therefore no difficulty should be experienced with ignition. The diffuser entry was 4 1/2 inches [112 mm] in diameter, and after a short parallel length it expended in two conical sections of total angles 4 degrees and 8 degrees respectively, to the diameter of the combustion chamber, 19 1/2 inches [500 m]. The exit diameter corresponding to a combustion chamber temperature of 1,000 degrees centigrade was 6.9 inches [175 mm]. The length of the combustion chamber was 47 1/4 inches [1,200 mm].

In the early experiments propane was used as the fuel on account of the ease with which it vaporizes. A change was later made to gasoline, and some experiments were also carried out with heavy diesel oil. The fuel was sprayed downstream from a ring containing a large number of orifices. The ring was backed by two cylindrical shields.

Tests were carried out on models over a range of combustion temperatures and air velocities. It was found that a compression efficiencies of 86 percent was obtained and an expansion efficiency of 92 percent. The velocity of air entrained over the surface of the duct was sufficient for adequate cooling. The tube temperature for

*Triebflügel Luftwaffe 1946...*the comic book. Courtesy: Press Antarctic Press.

established the principle. Subsequent work was directed towards reducing the dimensions of the equipment with a view to determining the optimum diameter size.

The diameter of the combustion chamber was progressive reduced to 250mm [9.8 inches], the entry and exit orifices remaining unaltered. At this stage the air velocity at the entry to the combustion chamber was about 100 meters/second [224 miles/hour] for air speeds of the order of 300 meters/second [671 miles/hour]. This high speed led to combustion difficulties. One serious difficulty with speeds of the order of 80-100 meters/second [179-224 miles/

At higher speeds, if the flame was blown out, burning would not recommence. To obtain a flame which would not blow out, an arrangement was devised consisting of a piece of porous material on to which gasoline was injected. A rich mixture was formed behind the porous material and this was ignited by a spark plug to give a steady flame. This gave satisfactory results up to the maximum speed.

At this stage models were tested in the wind tunnel of the *LFA* at Braunschweig for determination of the net thrust. An outer skin (cowling) was fitted over the diffuser to give a good aerodynamic shape, and it was found that the drag was rather less than half the

A tail sitting VTOL aircraft, *Focke-Wulf Flugzeugbau's Triebflügeljäger* would have utilized three un-tapered variable incidence rotor blades fixed to a rotary collar located about one-third of the way down the fuselage from the nose. They were to rotate around the fuselage, transmitting little or no torque to the fuselage.

Left: "Are you ready for this? The war is already lost and they are still creating new weapons!" *Triebflügel 1946.* **Courtesy: Press Antarctic Press.**

Two types of fighter/interceptors...in the foreground the *Focke-Wulf Fw Triebflügeljäger*, and in the background a *Heinkel He P.1078A* single seat fighter. Color computer generated digital image by *Gareth Hector*.

thrust. At 280 meters/second [626 miles/hour] airspeed the thrust recorded on the static test bed was 105 kilograms [231 pounds]. The corresponding net thrust in the wind tunnel was 58-60 kilograms [128-132 pounds]. The total external drag, calculated as the difference between these two thrusts, was about 25 percent, higher than the calculated skin friction of the outside surface.

A simple parallel section was not the optimum for the combustion chamber, from the point of view of internal flow. To maintain a constant pressure, which give optimum conditions, requires an expanding cross-section to allow for the contribution of combustion. This form was found to fit in with aerodynamic considerations on the external flow. *Dr. Zobel* of *LFA* Volkenrode near Braunschweig, working with *Professor Adolf Busemann* on aircraft forms, discovered that at speeds approaching the sonic region the best drag shape was an inverted drop with the maximum cross-section in the region

of 1/3 of the length from the tail. It was stated that at speed of 260-280 meters/second [581-626 miles/hour] this shape had only about half the drag of a drop shape with maximum cross-section at 1/3 of the length from the nose. The duct shape was therefore, altered so that the diameter was greatest toward the tail. The increase in net thrust was some 5-6 percent.

This last modification had led to some reduction in the length of the combustion chamber, as combustion was completed earlier through the maintenance of constant pressure. To obtain a further reduction in length experiments were initiated with injection of fuel forward, so that mixing should start in the diffuser. To facilitate mixing, the direction of the jet was skew to the duct to set up turbulence, and the fuel was ignited behind these baffles. Tests with different burner length of the combustion chamber to 700mm [2 1/3 feet].

A *Triebflügeljäger* alongside one of its brothers...the *Focke-Wulf Fw 190*. Scratch built fine scale models and photograph by *Günter Sengfelder*.

Otto Pabst discovered that the diffuser length could be reduced to 350/400 mm [14-16 inches], without serious loss of compression. A test with zero length diffuser, for example, a plain orifice, gave a 20 percent loss. Both *Sänger* and *Pabst* carried out with various forms of tail shapes in attempts to reduce drag. One thought, for example, was the sucking away of the boundary layer. The results were of doubtful value, as the differences recorded were only of the same order of magnitude as the error of measurement.

Commentary

The proposed *Triebflügeljäger* of 1944 was about as close as possible to fulfilling *Thomas Alva Edison's* sentiment that an airplane should be able to duplicate the maneuvers of a common humming-bird. It appears that a great deal of German scientific manpower was being applied to the ramjet technology and its use as a power plant for the *Triebflügeljäger*. Although the ramjet had potential, it is this author's personal opinion that the *Triebflügeljäger* was just too far out on the horizon of probability to be taken seriously. Nevertheless, *Oberst Siegfried Knemeyer* supported its research and development hoping for some major breakthrough. He was pushing the state-of-the art. A great deal of wind tunnel time was being given over to the testing small diameter and short length ramjets designed by *Dr. Otto Pabst* of *Focke-Wulf* and his colleague *Dr. Theodor Zobel* at *LFA's* A2 wind tunnel at Volkenrode near Braunschweig. So the *Triebflügeljäger projekte* was not some eso-

teric experiment, but a serious attempt at creating a production prototype VTOL. *Knemeyer, von Holst, Multhopp, von Halem, Mittelhüber, Pabst, Muck, Longner* and others would have liked nothing better than to have their *Triebflügeljäger* accomplish what the *Hawker-Siddely Aviation's Harrier* "Jump Jet" was able to achieve on its first test flight a mere sixteen years later on September 12th 1961. The *Harrier P1127* prototype on its first test flight lifted smoothly off in an unwavering vertical ascent from the thrust exhaust from its four nozzles. Moments later the flying machine started moving forward from its hover. It gathered speed in a surge of acceleration and streaked off into conventional flight.

As we have seen, a great number of VTOL designs have been produced and tried. The tail-sitting types include the *Triebflügeljäger, Lockheed's XFV-1, Convair's XFY-1*, and *Zborowski's SNECMA C.450.01 Coléoptère*. A few of these appear in many ways similar to a conventional airplane. Each of the above had a special tail on which it took off and landed. They were all abandoned, with the exception of the proposed *Triebflügeljäger*, because of the difficulty in maintaining fine control during landing when the fuselage was positioned vertically. A large number of "convertiplanes" are VTOL flying machines which can also fly horizontally with the same effectiveness as a conventional airplane. Some "convertiplanes" are conventional-looking aircraft which can tilt their rotors, or oversized propellers, so that the rotor's axis are vertical during take off and landing and are horizontal during for-

A port side view of the *Triebflügeljäger* featuring its very streamlined windscreen and cockpit canopy. Computer generated digital image by *Gareth Hector*.

spite the absence of airports or runways.

And what of the *Triebflügeljäger*? We can only speculate but it appears that this ramjet powered tail sitter would have suffered the same fate at the *Lockheed XFV-1*, *Convair XFY-1*, and *Count Zborowski's SNECMA C.450.01 Coléoptère*. They include:

• *Lockheed's Kelly Johnson* called the *XFV-1* the most riskiest military aircraft *Lockheed* ever designed and built. According to *Stephen Wilkinson, Kelly Johnson* said with the *XFV-1* "we couldn't look back over our shoulder when flying the thing and judge height. We could practice landing on clouds all day, but this is the only airplane we ever built which we were afraid to fly ourselves in the final tests."

• Where is the ground? The *Treibflügeljäger* pilot would have suffered the same difficulties seeing the ground during vertical landing as did the pilots of the *Lockheed XFV-1* and the *Convair XFY-1*. It would mean that the *Triebflügeljäger* pilot would have had to judge his altitude, attitude, rate of descent, and position over the ground while lying on his back and looking over his shoulder as he worked the throttle and flight controls to set up a controlled descent. *Convair* engineers said that to expect a pilot to be able to look down back over his shoulder and maneuver and land the *XFY-*

been very quickly accomplished.

• Prop wash coming its three ramjet powered rotors during vertical landing would have severely buffeted the *Triebflügeljäger* during its vertical landing, throwing up huge clouds of dirt and dust if it were not be put down on a hard surface such as concrete. Prop wash experience from the *Convair XFY-1* battered the flying machine around by its own turbulence...a rotating whirlwind of down wash;

• It was believed that basic training for a *Convair XFY-1* pilot would have required 50 hours in the Moffett Field (the former blimp hangar at San Francisco) tether. *Convair* experts including the *XFY-1's* only test pilot *James F. Coleman* believed that it should have been flown in a prone position. This way, as the pilot was attempting a vertical landing it would be a lot easier to look around to judge the ground.

• It is unclear if the *Triebflügeljäger's* three rotor blades, as some aviation historians have suggested, hinged, twisted, and flexed simultaneously acting as lifting and thrusting surfaces in forward flight. Drawings from *Focke-Wulf Flugzeugbau* suggest that the *Triebflügeljäger's* rotors would have not rotated during forward flight thereby acting as lift surfaces.

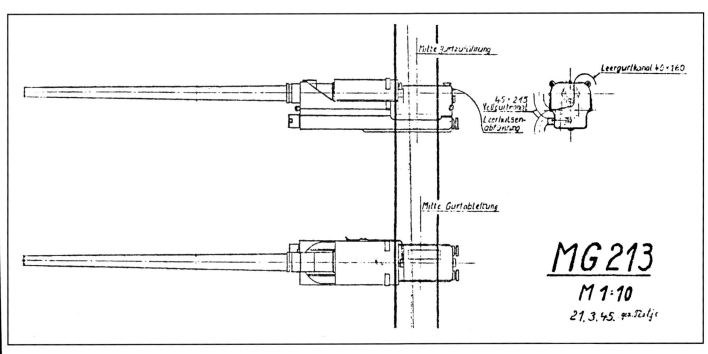

The *Triebflügeljäger* was reported to have been equipped with four *MG 213* "*Mauser*" *20 mm* cannon. Two *MG 213* would have been mounted on each side of the cockpit.

• It is unclear how the *Triebflügeljäger* would have transitioned from a helicopter to an airplane by simply pushing over from the vertical ascent to conventional horizontal flight.

• There appears to be a limit to the amount of speed and lift one can obtain with the help of a rotor. For example, if the speed of the rotor tips is 350 miles/hour and the forward speed of the *Triebflügeljäger* itself reaches 350 miles/hour, it is obvious that the actual speed of its rotor tips as they move forward would be 700 miles/hour. Conventional thinking states that the *Triebflügeljäger* would be in trouble, because its rotor tips are getting into the sound barrier. As a result the rotors would likely experience abnormal stress and vibration on the rotors as well as on the entire rotor system.

• It is unclear how the *Triebflügeljäger's* pilot would have coped with the howling thrust of its three (30 foot long) rotors as they revolved around the fuselage only inches aft the cockpit. In addition, the noise presented by the three small diameter short length

Otto Pabst ramjet engines would have been deafening to pilot and ground crew alike.

• It is unclear how the *Triebflügeljäger's* pilot would have psychologically handled the appearance "ring of fire" created by the operating of its three rotating small diameter short length *Otto Pabst* ramjet engines. This tell-tale signature of the *Triebflügeljäger* would have been even more spectacular at dawn or dust providing a warning to Allied fighter pilots and bomber crews alike.

• It appears that the *Triebflügeljäger*, as a point defense interceptor concept like its cousin the vertically launched bi-fuel liquid rocket powered *Bachem "Natter"* and the *Messerschmitt Me 163* were concepts which had major terminal flaws. They were like a shell from a well-aimed *Flak* cannon rising up vertically to hit bomber formations with minimal range and loitering ability.

• Returning to its base and a completed mission, the *Triebflügeljäger* had to again go vertical...transition from high-speed horizontal flight

The approximate locations on each side of the cockpit for the *MG 213 20mm "Mauser"* cannon. Notice that there is a slight stagger between the two cannon.

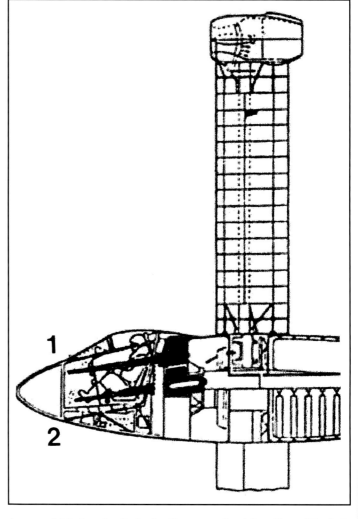

A pen and ink drawing featuring the approximate locations of the twin *MG 213 "Mauser" 20 mm* cannon port and starboard the cockpit.

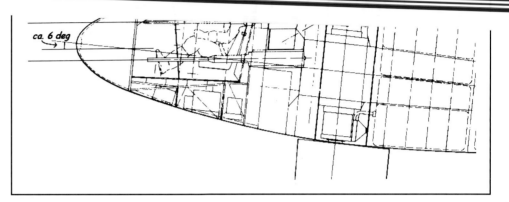

Design drawings of the *Triebflügel*, such as this pen and ink drawing featuring the location of the *MG 213 "Mauser" 20mm* cannon, show that they were positioned approximately 6 degrees below the VTOL's axis during forward flight.

ca. 6 deg

to a nose up hover...and then reverse course, backing down to a landing which meant that the pilot had no idea of where the ground was he struggled to look over his shoulder as he worked the ramjet throttle and flight controls.

• It is likely that the *Triebflügeljäger's* pilot would have been faced with three different configuration changes at the same time when the flying machine was pulled up into the vertical. These would have included trim change, power change, and taking your attention away from the instruments and trying to look back over your shoulder all the time trying to figure out when to add power since he would not have wanted to zoom backup too high because he'd have the difficult job of backing down again.

It is reported that the *Triebflügel* would have had the same cockpit interior and instruments as those found in the cockpit of the *Focke-Wulf Fw 190* fighter. Featured in this photo is the instrument panel of the *Fw 190*.

It is reported that the *Triebflügel* would have had the same cockpit interior and instruments as that of the *Focke-Wulf Fw 190* fighter. Featured is the port side cockpit panel of an *Focke-Wulf Fw 190*.

The starboard side cockpit panel of an *Focke-Wulf Fw 190*.

• Experience from the *Convair XFY-1* indicated, according to *Stephen Wilkinson*, that a descent rate greater than 10 feet/second found that the flying machine tended to tumble, totally out of control and at this altitude too low for a successful pilot ejection. The *XFY-1* when transitioning from horizontal to vertical flight was found to be inherently unstable. The same for *Lockheed's XFV-1*. According to the *XFV-1's* test pilot *Herman "Fish" Salmon*, "as you would slow down (for a vertical descent) and pull up about the time you hit 30 degrees and 60 miles/hour, she'd want to start to roll and be quite unmanageable. At that point the propwash was being deflected (due to the extreme angle of attack) and you weren't flying on the wing anymore, and the airplane was really stalled. Things would get progressively worse until you came to about 80 degrees, almost vertical, and then it would become quite easy to handle again."

• Experience from the *Convair XFY-1* showed that standing still at 500 to 600 feet off the ground the pilot lost his depth perception.

• The *Triebflügeljäger* would not have been a dog fighter. It was designed to be a point intercepter.

• It is unclear if the *Triebflügeljäger* ramjet engines at its three rotor tips had the ability for vectored thrust as some aviation historians and artists have described. The few drawings which exist of the *Triebflügeljäger's* rotor and its proposed small diameter *Pabst* ramjet engines do not indicate that the ramjets had the ability to be vectored. Nor do the drawings show the rotors containing any wing

A computer generated digital image by *Gareth Hector* featuring the *Triebflügeljäger* and its four fin-mounted outrigger wheels and wheel covers and its fuselage-mounted main wheel.

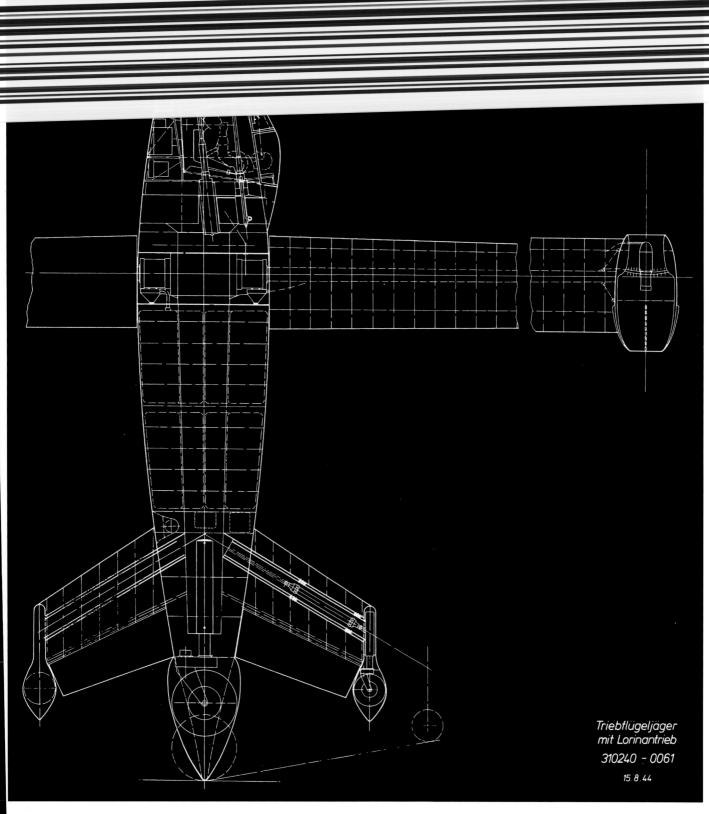

*Triebflügeljäger
mit Lorinantrieb
310240 - 0061
15.8.44*

A reverse negative featuring a pen and ink drawing by *Günter Sengfelder* of the *Triebflügeljäger* and its internal structure as viewed from its port side, and based on drawings from *Focke-Wulf Flugzeugbau* dated August 15th 1944.

twist or washout toward the tip. Perhaps a great many historians and artists have taken considerable liberty in presenting this proposed point interceptor. With its mission completed, the *Triebflügeljäger* would just as quickly returned to its base.

• There is a huge challenge in making a particular ramjet suitable for operations at different altitudes and speeds and requires unusual control mechanisms and geometries which were largely unknown and undeveloped in the mid 1940s. Thus any specific ramjet design coming from *Eugen Sänger, Otto Pabst, Helmuth Walter* would have been efficient only over a very narrow range of operating conditions. The *Triebflügeljäger's* ramjet engines would have severely limited its operational versatility.

• Unknown to *Eugen Sänger, Otto Pabst*, and *Helmuth Walter* at the time, was the problem of maintaining ramjet efficiency at all

altitudes of operation. Post war ramjet research would lead, for example, to ramjet designs in which oxygen bearing compounds were carried in addition to normal fuel (usually kerosene) and with an automatic means of supplying of oxidant (oxygen) to keep the ramjet's flame front stationary.

• An operational ramjet engine requires an automatic fuel metering control system. *Eugen Sänger, Otto Pabst, Helmuth Walter,* and other ramjet researchers had yet to develop automatic fuel regulators, fuel metering, and flame "blow-out" protection. None of these essentials had been perfected at the time (September 1944) the *Triebflügeljäger's* overall design had been completed by *Focke-Wulf's Flugbaumeister Heinz von Halem* with help from *Ludwig Mittelhüber*. It appears that *Erich von Holst's* ramjet powered *Triebflügeljäger's* fuselage could have been easily built, however,

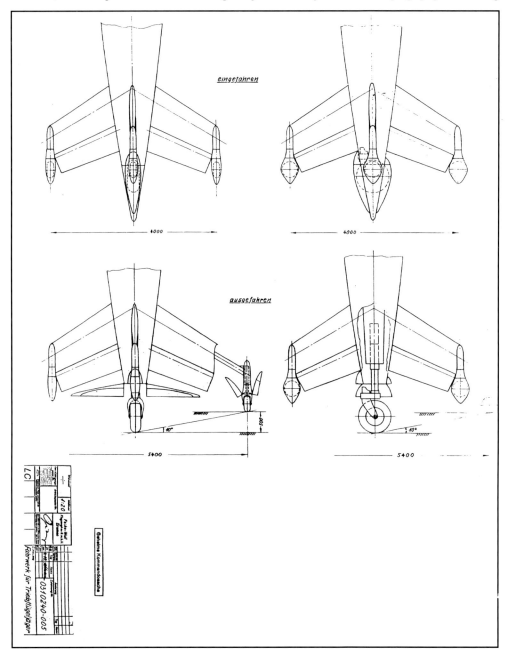

A drawing by *Günter Sengfelder* **of the** *Triebflügeljäger's* **four outrigger wheels and single main wheel based on drawings from** *Focke-Wulf Flugzeugbau* **dated August 4th 1944.**

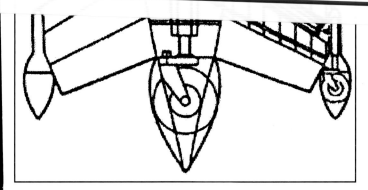

A close up drawing of the *Triebflügeljäger's* single main wheel.

it would have probably been placed in storage waiting for its ramjet engines which would have been several decades away.

• Fuel economy and range. Ramjet engines consume enormous quantities of fuel at low Mach numbers. This means that the three small diameter *Otto Pabst* ramjets selected by *Hans Multhopp* to power

you try to do it in real life.

• The *Harrier* in real life is not a VTOL but a STOVL, that is short take off and vertical landing. They had to be overloaded for a military operation and then they were no longer able to rise vertically but become what is called "flat risers" because in order to get airborne they would have to run down a short runway or the deck of an aircraft carrier before the combination of wing lift and downward vectored jet exhaust thrust lifted it aloft. A vertical take off *Harrier* came only at air shows when the flying machine was light and without its usual load of drop tanks, bombs, and missiles...hardly a combat ready condition.

• Regarding the *Lockheed XFV-1* and the *Convair XFY-1*, said author *Stephen Wilkinson*, test pilot *James F. Coleman* of *Convair* "was one of the last people ever to venture aloft in a machine that nobody how to fly, that no simulator had proved would fly, and that no computer could promise would be controllable." The same can be said of *Focke-Wulf Flugzeugbau's* proposed point interceptor fighter...the *Triebflügeljäger*.

A close up look at one of the *Triebflügeljäger's* four outrigger wheels and its clamshell style wheel cover.

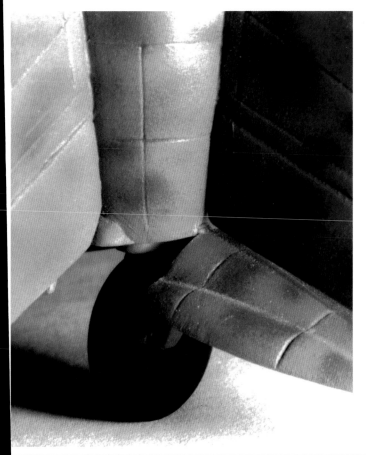

Left: A close up look at the *Triebflügeljäger's* single main wheel and its clamshell style wheel cover.

The outrigger wheels with their clamshell covers appear rather delicate structures, as shown on this scratch built fine scale model photo by *Günter Sengfelder*. It would appear that the *Triebflügel's* pilot would have to land this VTOL level, touching the landing pad on all four outrigger wheels.

It would appear that if the *Triebflügel* did not land absolutely perpendicular to the ground the outrigger wheel and its support could easily collapse, causing this VTOL to tip over. Scratch built fine scale model and photograph by *Günter Sengfelder*.

Ronnie Olsthoorn

Some final thoughts on how the *Triebflügeljäger* might have behaved in horizontal flight and if its three rotors would have been stopped for more efficient straight flight. *Ronnie Olsthoorn*, whose digital images appear in this book, also holds a degree in aeronautical engineering. He believes that the *Triebflügeljäger's* three rotors would have continued to rotate during flight based on several factors such as:

(1) Wing Warpage: The design drawings which have survived of the *Triebflügeljäger's* rotors show them warped like a propeller blade. This shape is ideal for a propeller or rotor and provides a good lift distribution across the entire wing span. A high angle of pitch at its attachment base (inside) where the air speed is slow. A low angle of pitch towards its tip (outside), where the air speed is high due to rotation. Thus the rotors would appear to have optimum lift throughout its entire span. In forward flight the rotor's pitch is reduced to keep the optimum, as the rotation speed is combined with the forward speed. On the other hand, wing warpage makes it very difficult to obtain a good lift distribution in horizontal or straight flight. However, on the twin rotor version of the *Triebflügeljäger* there appears to be no wing warpage. Instead, its rotors are constructed straight which means that they have been made for straight flight. Vertical take off could have been achieved by rotating the

Right: The one and only VTOL flying machine researched and developed by *Nazi* **Germany was their bi-fuel liquid rocket propelled** *Bachem Ba 349 "Natter"*, **shown here mounted on a bark stripped pine tree pole field launch rack. The ladder was needed in order for the pilot to mount this bi-fuel liquid rocket propelled target interceptor. Scratch built fine scale model and photograph by** *Jamie Davies*.

a huge 30 degree "V" angle.

(3) Ramjet Engine Position Angle: Each of the three ramjets are placed under an angle, due to the rotor's warpage, and it would be apparently very complicated, if not impossible, to assure the *Triebflügeljäger* a straight forward thrust this way.

(4) Technical Drawings Featuring Pilot Position and Cannon: On original drawings of the *Triebflügeljäger* they clearly show its four cannon position under an angle. Also the pilot looks down under that same angle. When one rotates the drawings so that the pilot looks straight ahead and the cannon all fire straight ahead as well, the angle is just over 6 degrees. This appears to be the attitude of the *Triebflügeljäger* in normal flight at cruise speed. Its rotors will still be rotating and producing thrust in the length of the fuselage's centerline. This thrust can be split up in forward thrust to give the flying machine forward airspeed and lift to keep it up in

Left: A *Bachem Ba 349 "Natter"* in a field launch position...in a metal framework attached to a stripped pine tree. This *"Natter"* is shown from its port side. *Erich Bachem* envisioned a simple 70 foot tall pine tree stripped of its bark once mass production of the *Bachem Ba 349's* got underway. Pine trees were to be found all over Germany...a 70 foot high pine tree could be cut down, stripped of its bark, and quickly converted into a launcher for a *"Natter."* Below: Color computer generated digital image by *Ronnie Olsthoorn*.

the air. I am convinced that the rotors of the *Triebflügeljäger* kept rotating during horizontal or straight or forward flight.

Furthermore, a major problem becomes apparent straight away. When the *Triebflügeljäger's* drops, the angle of the flying machine at which the its flies will become greater, thus its cannon will not be pointing straight ahead any longer. In addition, due to the *Triebflügeljäger's* rotating rotor, the maximum speed of this flying machine would not be very impressive. A rotor/propeller creates a fair amount of drag. Given the straight rotor design, the *Triebflügeljäger* is unlikely to speed along faster than 435 miles/ hour [700 kilometers/hour] perhaps maybe never more than 373 miles/hour [600 kilometers/hour]. Nowhere near Mach 0.9 [667 miles/hour], which has been suggested in several historical publications.

The only apparent advantage of the *Triebflügeljäger* would have been its great agility. I am confident that the *Triebflügeljäger* would have been able to turn on a dime. All it would have been required was a mere touch of the tail's control surfaces and it would have turned around. Look at it as the equivalent of a propeller on a stick, as first invented by the Chinese thousands of years ago. Many youngsters have played with such a thing when a child. I seen film footage of all the early VTOL such as the *Lockheed XFY-1 "Salmon"* and the *Convair XFV-1 "Pogo."* Even the fairly conventional *"Pogo"* was quite a complicated machine to fly and almost impossible to vertically land. The *Focke-Wulf "Triebflügeljäger"* would have been much worse, I suspect.

In short, I believe that the *Triebflügeljäger* would have been technically achievable, but operationally a total failure. Flying this machine may have been extremely dangerous. Aiming its cannon in a dogfight and expecting to hit a moving target in all likelihood would have been next to impossible.

Bibliography

• Erich von Holst, *Vom Rätsel des Vogelfluges*, Modellgleitflug-Post, September/Oktober 1948, page 10;

• Erich von Holst, *Vom Rätsel des Vogelfluges*, Modellgleitflug-Post, November/Dezember 1948, page 18;

• Erich von Holst, *The Behavioural Physiology of Animals and Man*, University of Miami Press, Coral Gables, Florida, 1973;

• Marti Sarigul-Klijn, *I Survived the Rotary Rocket,* Air & Space Smithsonian, March 2002.

• Dietrich Küchemann & Johanna Weber, *Aerodynamics Of Propulsion*, McGraw-Hill Publishing Company, London, pages 248-260;

• H.J. Meier, *Das Kuckucks ei*, Modellflug-Post, April/Juni 1949, pages 98-100;

• Justo Miranda, *Dossier #9 V.T.O. Interceptor,* The Reichdreams Research Services, Madrid, Spain, no date;

• Bruce Myles, *Jump Jet: The Revolutionary V/STOL Fighter*, Presidio Press, San Rafael, California, 1978;

• Otto E. Pabst, *Die deutsche Luftfahrt Kurzstarter und Senkrechntstarter,* Bernard & Graefe Verlag, Koblenz, 1984;

• Stephen Wilkinson, *This End Up: The Rise and Fall of the Pogo Fighters,* Air & Space Magazine, 1996;

Color computer generated digital image by *Ronnie Olsthoorn*.

Color computer generated digital image by *Ronnie Olsthoorn*.

Color computer generated digital image by *Ronnie Olsthoorn*.

Color computer generated digital image by *Gareth Hector*.

Color computer generated digital image by *Gareth Hector*.

Color computer generated digital image by *Gareth Hector.*

Color computer generated digital image by *Gareth Hector*.

Color computer generated digital image by *Gareth Hector.*

Color computer generated digital image by *Gareth Hector*.

Color computer generated digital image by *Gareth Hector.*

Color computer generated digital image by *Gareth Hector*.

Color computer generated digital image by Gareth Hector.

Color computer generated digital image by *Gareth Hector*.

Color computer generated digital image by *Gareth Hector*.

Color computer generated digital image by *Marek Rys.*

Color computer generated digital image by *Marek Rys*.

Scale model and color photograph by *Dan Johnson.*

Scale model and color photograph by *Dan Johnson.*

Scale model and color photograph by *Mark Hernandez*.

Scratch built fine scale model and color photograph by *Günter Sengfelder.*

"Yellow 4." Digital image by Ronnie Olsthoorn.

"White 15" featuring a white **"lighting bolt"** fuselage nose emblem. Digital image by Ronnie Olsthoorn.

"Yellow 4." Digital image by Ronnie Olsthoorn.

Right: "Yellow 4" lifting above its forest hiding place. Digital image by Ronnie Olsthoorn. Below: As many as four Triebflügels standing on the tarmac in perfect alignment dimly seen in the early morning heavy fog. Digital image by Jozef Gatial.

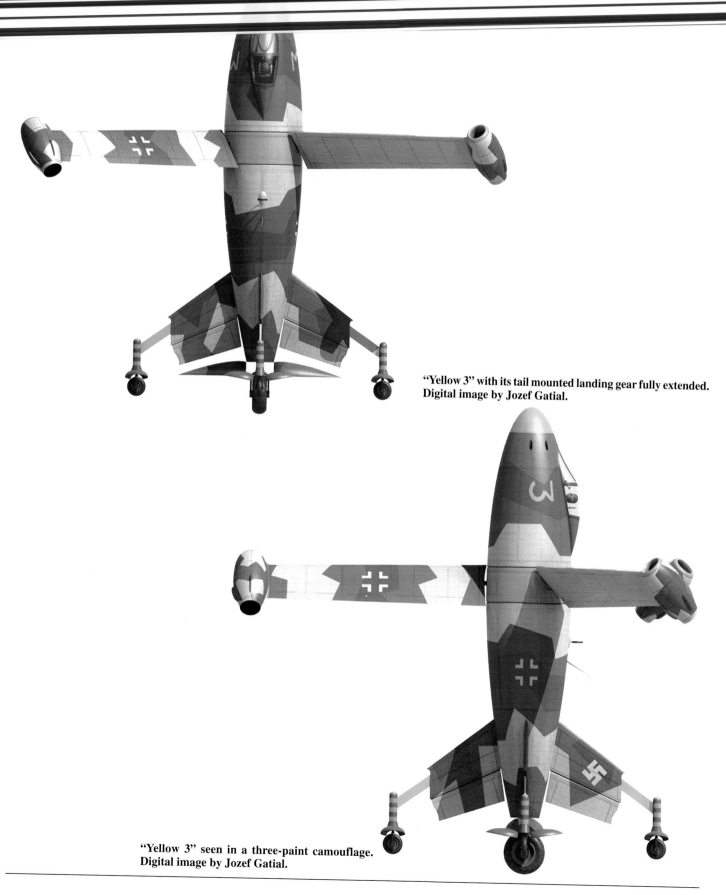

"Yellow 3" with its tail mounted landing gear fully extended.
Digital image by Jozef Gatial.

"Yellow 3" seen in a three-paint camouflage.
Digital image by Jozef Gatial.

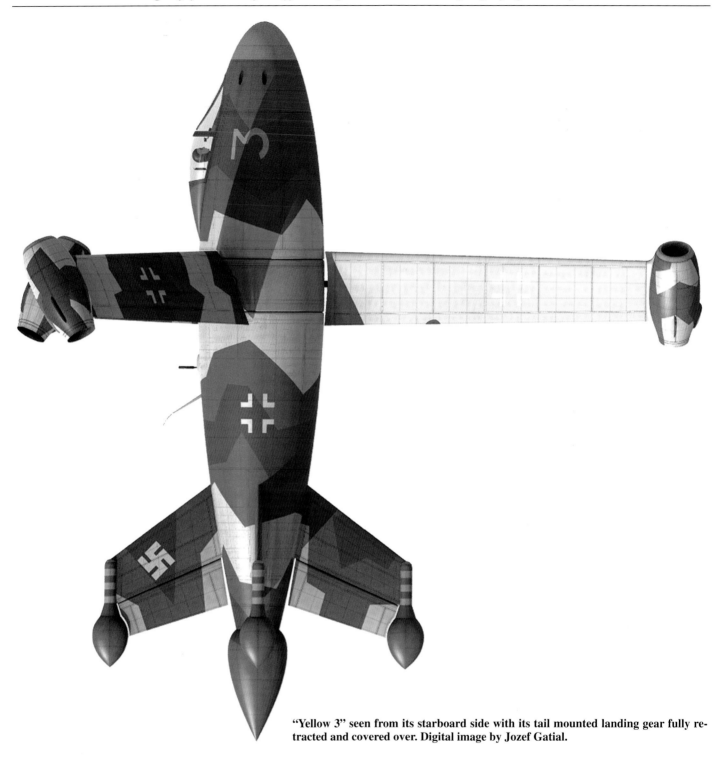

"Yellow 3" seen from its starboard side with its tail mounted landing gear fully retracted and covered over. Digital image by Jozef Gatial.

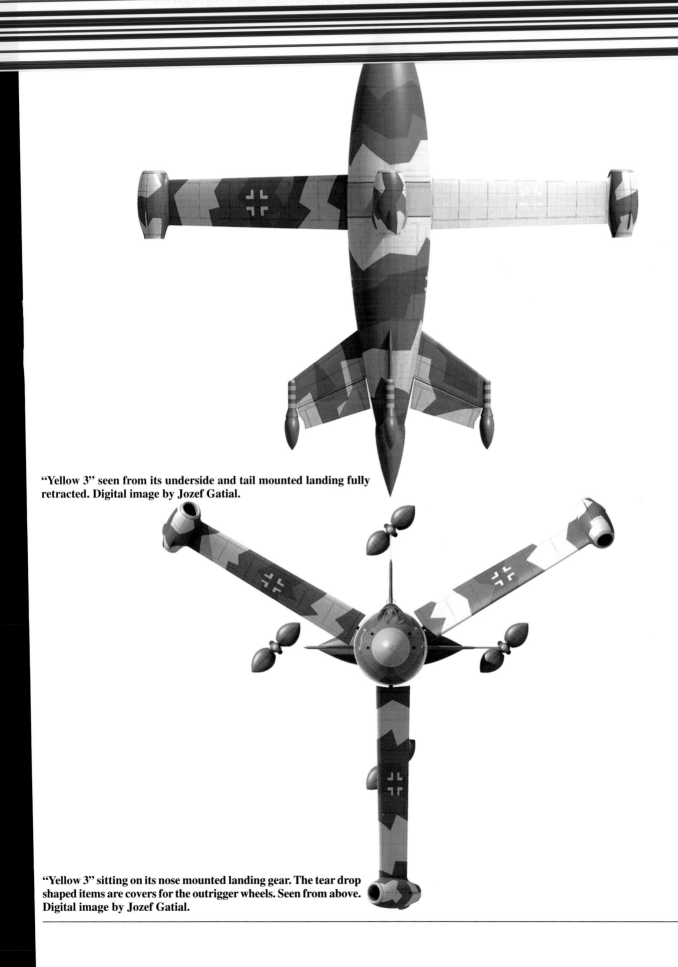

"Yellow 3" seen from its underside and tail mounted landing fully retracted. Digital image by Jozef Gatial.

"Yellow 3" sitting on its nose mounted landing gear. The tear drop shaped items are covers for the outrigger wheels. Seen from above. Digital image by Jozef Gatial.

"Yellow 3" seen from beneath featuring its tail mounted landing gear fully extended. Digital image by Jozef Gatial.

"White 9" covered overall in a single RLM dark gray camouflage. Featured is its full length cockpit and tail mounted landing gear fully retracted. Digital image by Jozef Gatial.

All RLM dark gray camouflaged "White 9" featuring its "bumble bee" fuselage nose emblem. Digital image by Jozef Gatial.

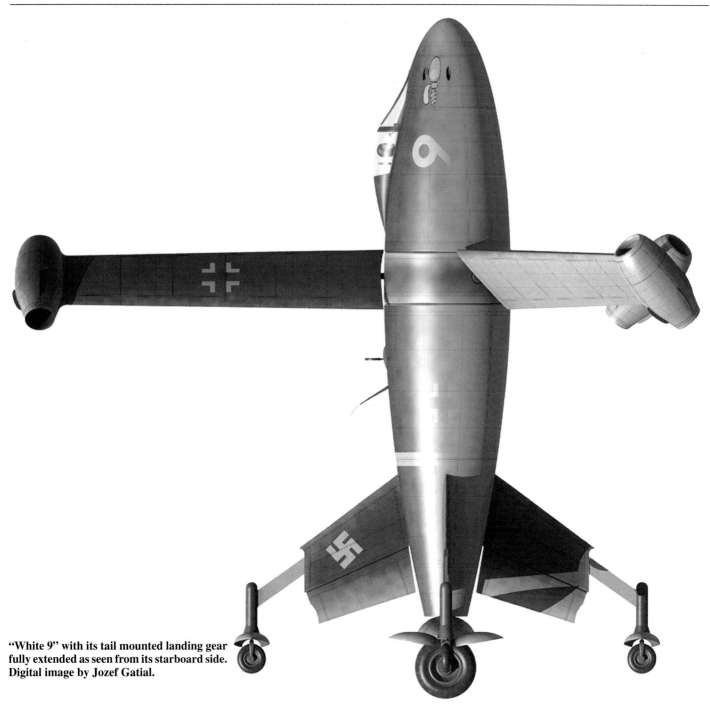

"White 9" with its tail mounted landing gear fully extended as seen from its starboard side. Digital image by Jozef Gatial.

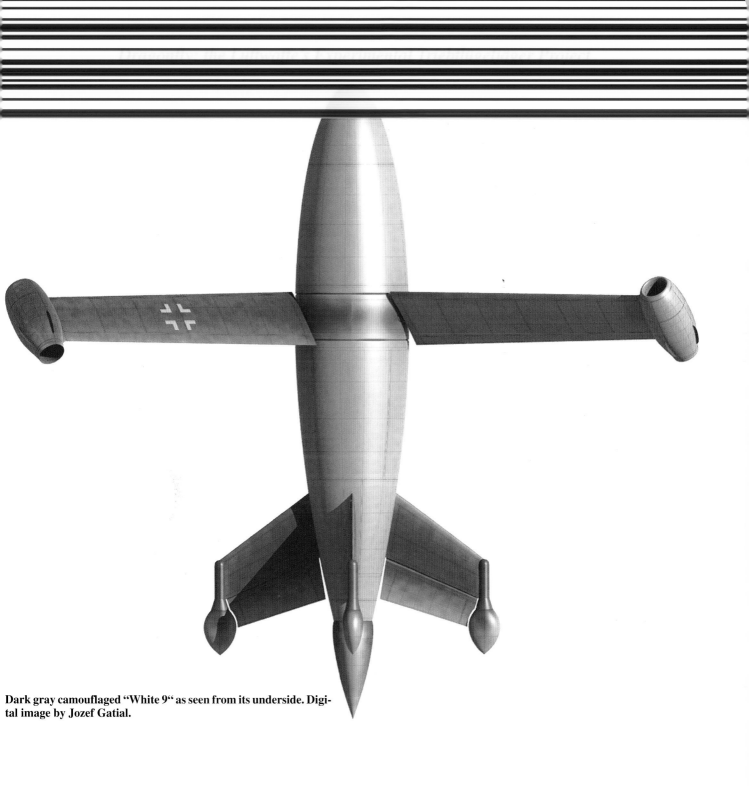

Dark gray camouflaged "White 9" as seen from its underside. Digital image by Jozef Gatial.

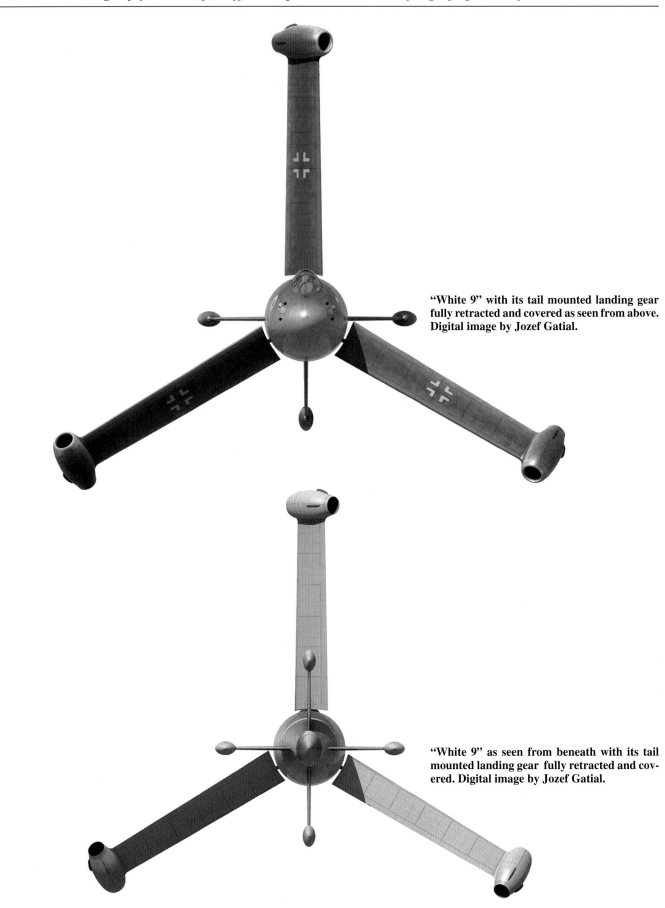

"White 9" with its tail mounted landing gear fully retracted and covered as seen from above. Digital image by Jozef Gatial.

"White 9" as seen from beneath with its tail mounted landing gear fully retracted and covered. Digital image by Jozef Gatial.

"White 4" seen in a rare scribble patterned camouflage as seen from its nose port side. Digital image by Ronnie Olsthoorn.

Above: A pair of Triebflügels [Yellow 3 left and Yellow 4 right] with pilots in the cockpits seen at dawn. Below: With the fog dissipated "White 6" is airborne, while "White 9" is making ready for lift off. Digital images by Jozef Gatial.

Above: "Bumble bee" emblemed all dark gray "White 6" [right] and "White 9" [left] are in lift off together from their forest hideaway. Below: "Yellow 4" [right] and "Yellow 3" [left] in their tri-color camouflage are seen lifting off from their forest hideaway at dawn. Digital images by Jozef Gatial.

"Yellow 3" [right] and "Yellow 4" [left] seen shortly after lift off and before their tail mounted landing gear has been retracted. Digital image by Jozef Gatial.

"Yellow 3" [left] and "Yellow 4" [right] are seen in full flight with their rotors in maximum rotation and tail mounted gear fully retracted as they attempt to gain altitude in pursuit of high-flying American Boeing B-17 bombers. Digital image by Jozef Gatial.

Two all gray camouflaged Triebflügels "White 9" [right] and "White 6" [left] seen from beneath after lift off and striving for high altitude as fast as possible. Digital image by Jozef Gatial.

Above: A nose-on view of all gray "White 9" [left] and "White 6" [right] proceeding at maximum rotor rotation. Below: All gray "White 9" [left] and "White 6" [right] seen from their tail port sides. Digital images by Jozef Gatial.

Above: A pair of tri-color camouflaged Triebflügels, "Yellow 3" [left] and "Yellow 4" [right], seen gaining altitude after leaving their forest hideaway. Digital image by Jozef Gatial. Below: "Yellow 3" [left] appears to have fixed its rotors, turning them into airfoils to achieve maximum forward speed. It is unclear how this transition would have been accomplished. "Yellow 4's" [right] rotors continue to rotate while the pilot goes about making the transition, too. Digital images by Jozef Gatial.

Above: All gray "White 9" [left] and "White 6" [right] are seen striving for altitude against a background of a blue/white snow covered mountain range. Below: "White 6" [left] and "White 9" [right] are seen from their port side with a blue/white snow covered mountain range in the background. Digital images by Jozef Gatial.

Above: "White 9" [left] and "White 6" [right], each with their rotors fixed, are closing in on their next victim...a Boeing B-29. Below: Tri-color camouflaged "Yellow 4" [left] and "Yellow 3" [right] as seen from their nose port sides. Digital images by Jozef Gatial.

Above: "Yellow 3" [left] and "Yellow 4" [right]. Below: Tri-color camouflaged "Yellow 3" [left] and "Yellow 4" [right] seen at early dawn under full rotation power. Digital images by Jozef Gatial.

"Yellow 4" [left] and "Yellow 3" [right] are seen closing in on their prey at dawn...a Boeing B-29. Digital image by Jozef Gatial.

Two all gray Triebflügels gaining altitude at dawn after lift off from their hideaway in a mountainous area in southern Germany. Digital image by Jozef Gatial.

Two all gray Triebflügels..."White 9" [left] and "White 6" [right], as seen from their tail port sides. Digital image by Jozef Gatial.

A close up of the aerodynamic pleasing Triebflügel "White 9" of "bumble bee" squadron and seen from its nose port side. Digital image by Jozef Gatial.

Two Boeing B-29s [the prey] and two Triebflügels [the hunters], each featuring "bumble bee" fuselage nose emblems. Digital image by Jozef Gatial.

"White 6" [left] and "White 9" [right] are seen returning to their base hideaway, and are in the process of making a transition to a vertical landing. Digital image by Jozef Gatial.

Mission accomplished. "White 6" [upper] and "White 9" [lower] immediately return to their hideaway, and are seen moments away from touch down. Digital image by Jozef Gatial.